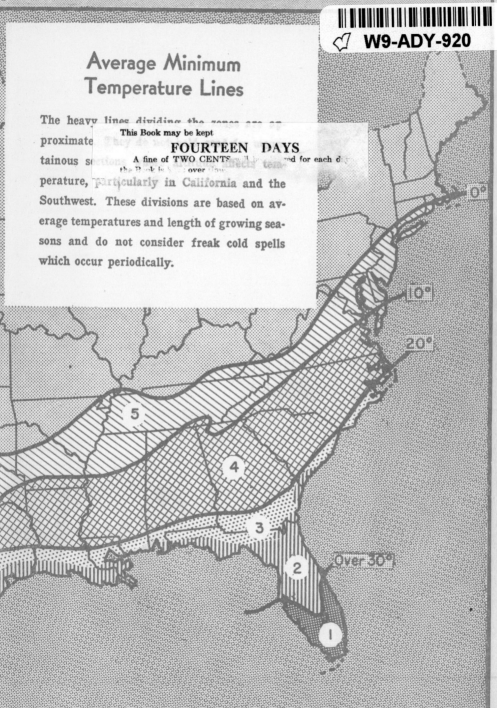

Average Minimum Temperature Lines

The heavy lines dividing the zones are approximate. They do ~~~~~ ~~~~~ ~~~~~
tainous s~~~~~ ~~~~~ ~~~~~ affects temperature, particularly in California and the
Southwest. These divisions are based on average temperatures and length of growing seasons and do not consider freak cold spells
which occur periodically.

W9-ADY-920

0°

0°

20°

5

4

3

2

Over 30°

1

FLOWERS OF THE SOUTH

FLOWERS
OF THE SOUTH

Native and Exotic

By WILHELMINA F. GREENE AND HUGO L. BLOMQUIST

THE UNIVERSITY OF NORTH CAROLINA PRESS

Chapel Hill

PREFACE

THE SOUTH has the richest flora in the United States, and yet little is known concerning it by the average individual. The few books which have been published on the plants of this area have been mostly too technical for anyone who is not a systematic botanist, and the majority of the population does not belong to this class of specialists. Still there are many who are interested in plants, both native and introduced, and who are anxious to know the names of plants, their natural history, and the uses they may serve for man's material needs and enjoyment. It is primarily for these and not for the technical botanist that this book has been prepared.

As is evident from the illustrations, this is to a great extent the artist's book. This is also true from the standpoint of its history. Twenty years ago, Mrs. Greene's flower sketches came to the attention of the late Mary Francis Baker, author of *Florida Wild Flowers,* who encouraged her to continue this work with the idea of illustrating a popular book. While teaching nature study and art in the public schools, she was impressed by the lack of published material on southern plants that could be used advantageously. Repeated requests from Mrs. Reasor, Garden Club Director of Junior Work, and from junior chairmen throughout the state of Florida, convinced her that a well-illustrated book in non-technical language was needed.

The preparation of the present text was commenced after most of the illustrations had been made. It was the responsibility of the writer of most of the descriptions, Dr. Blomquist, to suggest deletions and additions that would make the selection representative and useful. To select some five hundred plants out of nearly six thousand is obviously a task that can be questioned from many points of view—as can the inclusiveness of individual descriptions. Space limitations imposed by the necessity of producing a volume not beyond the means of the average user of such a book played a major part in all decisions. The selection has leaned more to the objective side than to that of the sole preference of the authors.

The sequential arrangement follows in general that of John Kunkel Small's *Manual of the Southeastern Flora* (1933) for native plants and that of L. H. and Ethel Zoe Bailey's *Manual of Cultivated Plants* (1949) for cultured exotics. However, for the sake of artistic arrangement of illustrations, certain minor deviations from a rigid systematic sequence have sometimes been found necessary. The technical names of native plants are in general those of *Gray's Manual of Botany,* M. L. Fernald, 8th ed. (1950), so far as distribution of plants allows, because of its broad concepts of families, genera, and species, as well as for being the most recent volume. For strictly southern plants, Small's *Manual* has been used. For cultivated exotics, the names are mostly according to Bailey's *Manual* and *Hortus* (1935). The selection of common names is always a difficult matter as these often change in time and with location. Most of them have been selected from the above manuals while others are sectional names which have not yet come into general use.

In the line drawings of native plants, the artist has emphasized the white flowers by the blackening of the stems and foliage. However, for cultivated exotics this technique is used simply for artistic effect. Where a flower occurs also in a color illustration, reference to this picture is often made in the text. In the use of colored mixed bouquets, the artist is anxious to have it understood that these groups do not represent floral arrangements. They are used mainly to show what flowers predominate in certain sections at some particular season. For those who may use this book for identification of plants, it is suggested that the line drawings be colored with water color pencils as the reader finds the flowers in nature. This is a good way to learn form and color of flowers as well as to recognize them upon a second encounter.

During the many years the artist has been drawing and painting plants, she has come in contact with many individuals who have in one way or another contributed valuable aid to her efforts. She is deeply indebted to the following for the identification of plants: Dr. Richard A. Howard, Dr. Edgar T. Wherry, Dr. Herman Kurz, Dr. Erdman West, Dr. Wilbur H. Duncan, and the late Mrs. Mary Francis Baker, Dr. John Kunkel Small, and Mr. William Buswell. For other botanical information and help, she wishes to express her sincere appreciation to the following: Dr. Woolford B. Baker, Dr. C. J. Th. Uphof, Mr. Ralph Wheeler, Dr. Robert B. Platt, Mr. John V. Watkins, Dr. Hamilton P. Traub, Dr. Roland M. Harper, Mr. Mulford B. Foster, Mr. B. Y. Morrison, Dr. H. H. Hume, Mr. Wyndham Hayward, Dr. J. B. McClelland, Dr. and Mrs. H. R. Totten, Dr. Marie Wagner Cobb, Miss Pauline Hall, and to Mrs. Kathryn Small Garber for permission to copy some of the enlarged drawings from the *Manual of the Southeastern Flora* by her father, Dr. John Kunkel Small. She wishes especially to thank Mr. M. J. Daetwyler for reading her manuscript on the exotics and for his suggestions as to zones; and her appreciation to Dr. Paul A. Vestal for his assistance and for allowing her to use the herbarium and laboratory at Rollins College. The artist wishes to thank her many friends in the Florida and other garden clubs, too numerous to mention here, for their help and encouragement through the years, especially those who have accompanied her on collecting trips. She is also indebted to all those who have permitted her to paint flowers from their gardens, especially to Mrs. J. Norman Henry of the Henry Foundation for Botanical Research, Mr. E. C. Robbins of the Gardens of the Blue Ridge, Miss Caroline Dormon, and Mrs. Herbert Smith.

W. F. G.
H. L. B.

CONTENTS

COLOR PLATES

INTRODUCTION

THE SOUTH ABOUNDS in many beautiful and interesting plants, and the ones which attract our attention most—and are indeed the most common—are the flowering plants. Our admiration of flowers has made the word "flowers" almost synonymous with the plants that bear them. Still, we often admire as much the plant's form, its foliage, and fruits. Our flowering plants range from tree giants, such as magnolias, the tulip-tree, and sweet-gum to the tiny *Bartonia* with a slender stem a few inches tall bearing microscopic leaves and terminated by a solitary flower. Besides native plants there are also many which have migrated from other parts of the world and become more or less naturalized or have been introduced for ornamental or agricultural purposes. Some of these may flourish in our gardens and fields where we do not want them and then we call them "weeds." Nevertheless, many of these have their distinctive beauty which we may fail to appreciate because they are waifs and "common."

To know and appreciate our native plants is important. It is to a certain extent an indication of an advanced civilization and culture. One of the first essentials in knowing plants is to know their names. Plant names are generally of two kinds, the technical or botanical, and the common or folk names. Both have a place in our language, but they have different origins and uses. The technical names are given to plants by systematic botanists; they must conform to certain rules and regulations and are cosmopolitan in their application. Common names, on the other hand, may be coined by anyone and vary from place to place and may even change in time. They usually allude to some characteristic of the plant, real or imaginary, or to its usefulness to man, actual or hypothetical. To most non-botanists, technical names have little meaning as they are latinized and may be of ancient origin with unknown meanings, descriptive, or derived from personal or geographical names. To call a plant *Myosotis scorpioides* means little if anything to the average individual, even when translated to "mouse-ear scorpion-like," but when it is called "forget-me-not" the meaning is clear and the plant becomes a symbol of sentiment. It would be better if more common names of plants were more widely used, as are technical names; and attempts have been made to standardize common names but it is doubtful to what extent this will ever be successful. To know plants requires, of course, more than knowing their correct botanical or common names. It demands an acquaintance with the plants as living things. This is essential in dealing with them successfully in the garden, in landscaping, as well as in their conservation and preservation. When the children of the Coastal Plain call

[xi]

pitcher-plants "cups-and-saucers" and trailing-arbutus "mountain-roses," they are not using technical or standardized botanical names, but it is apparent that they are familiar with the plants.

The use of more native plants for ornamental purposes in the southern states has great possibilities. They are as attractive and desirable, in some cases more so, as those found in other parts of the world. This fact was soon appreciated by the European horticulturalists during the early explorations of this country. The collecting of seeds and seedlings of our native plants and their export to Europe was an important aim of some of the early botanists in America, such as André Michaux of France, Thomas Walter and James Drummond of England, and even of some of our own botanists, such as John and William Bartram of Philadelphia. The South's colorful Rhododendrons have been cultivated in England for over a century and are prized there, as well as in other countries, more than they are in their native land. However, in an extensive use of our native plants for horticultural purposes, there is danger of reducing some of them in nature to the point of extinction, unless they are propagated in nurseries. This happened to franklinia, the lost-camellia of Georgia. Fortunately this beautiful tree, because of its wide adaptability and ease of propagation, has been preserved through cultivation and should be planted more widely than it is. Individuals who collect plants for their gardens from nature should do so sparingly and use more of seeds and cuttings than whole plants. Mrs. J. Norman Henry, an ardent conservationist, says, "I soon learned to be a good propagator, and I never take more than a 'start' of a plant, often preferring seeds, and frequently tag a plant when in bloom and then return for seeds in the autumn." Other individuals and many organizations, such as the North Carolina Wild Flower Preservation Society and the Lily Society are encouraging the planting of seeds of wild plants. The development of private wild flower gardens is a growing hobby which is a pleasant way of becoming familiar with wild plants, and it may contribute to their preservation locally where they are in danger of being destroyed by the clearing of land and road construction. Through the popularity of the wild flower garden, it has become common knowledge that many may be grown successfully far beyond their natural ranges. Another use of wild flowers, fruits, and foliage is in flower arrangements, which has wonderful possibilities for creating many new and unique designs and artistic effects.

The conservation of our native plants is important, but it is not a simple matter. It is of primary importance that we develop a conservative state of mind and spirit. Conservation should be, first of all, based upon a thorough knowledge of the plants concerned, which at present is, in many instances, far from complete. We should know more about the abundance, distribution, relation to soils, and reproduction of the plants concerned. It is unwise to generalize, for to make a strenuous effort to protect plants which are abundant, widely distributed, and reproduce freely is not necessary and a waste of effort, and may even lead to disadvantage to those plants which need protection. However, we must face the fact that as man increases and population becomes more dense in habitable areas, his activities make more or less serious inroads upon his environment. Vegetation is usually removed, and the soil is exposed to erosion and often transported from place to place by steam-shovels and bulldozers. Much of this is undoubtedly necessary for human convenience and welfare, but when carried out without the slightest consideration of what is in the environment and without regard for the future, it may become unnecessarily destructive. The same may be said of flooding extensive areas for the construction of dams for irrigation and electric power. In this way there is danger of destroying many valuable plants and exterminating rare ones. But these are not the only activities which lead to often unnecessary extermination of our native plants. One danger lies in the extensive collecting of plants for commercial purposes. Ginseng, which has for many years been collected and shipped to China where its high value is based upon superstitious notions, has as a result become so rare that it is very unusual to find a plant in its natural range. The same is happening to many so-called "drug plants" the uses of which often have no therapeutical value. Included among these is the beautiful pink lady's-slipper which, although fairly abundant where found, has been completely exterminated from the vicinity of many urban communities. Another example is the extensive collecting of the vanilla-plant or deer-tongue of our Coastal Plain for use in the flavoring of tobacco products. The use of large quantities of evergreens for Christmas decoration of city streets is rapidly diminishing our native hollies, running-cedar, and ground-pine. The collecting and direct sale of some of our native flowering plants should be discouraged. In this way the gorgeous yellow lady's-slipper has almost completely disappeared from the Piedmont areas. A few years ago, the Venus' flytrap of southeastern North Carolina was being dug up and sold in great quantities as a curiosity. Fortunately due to the conservative-minded individuals and the state's garden clubs this activity was prohibited by law. Similar steps will no doubt have to be taken in the future for the protection of other plants. If plants are to be used, they should be propagated as are many other economic plants or they will be wiped off the face of the earth.

An ideal way of preserving at least some part of our flora, is the establishment of certain areas as national, state, or municipal plant preserves. This has the great advantage of leaving the plants in their native habitats and communities. Such areas are of considerable educational value, and, when properly administered, contribute to wholesome recreation. Those who are interested in plants should use their influence in establishing more such areas in the southern states, as did the artist and her husband in the establishment of Highlands Hammock of Sebring, Fla., as a state plant preserve. Similarly, many organizations, such as garden clubs and wild flower preservation societies can do much in preserving our native flora. Several years ago, the garden club of Tryon, N.C. acquired through purchase a tract of 300 acres supporting a most remarkable assemblage of native plants of the southern Blue Ridge Mountains, which has become the famous Pearson's Falls Wild Flower Preserve. Individuals who are interested and are fortunate in having the means are also doing their part in plant preservation. A noteworthy effort of this kind is the private plant sanctuary of Mrs. J. Norman Henry of Gladwyne, Pa., known as the Henry Foundation for Botanical Research. Here Mrs. Henry has brought together by the expenditure of her own funds and physical effort numerous rare and beautiful wild flowers from all over North America. For over 20 years she has distributed free of charge seeds and seedlings of rare plants to botanical gardens in various parts of the world and has placed many with commercial growers in this country.

The South is noted for its beautiful home gardens as well as the larger gardens of some of the coastal communities of the southeastern states. They show what can be done with exotic Camellias, Azaleas, Hibiscus, and many others. But a garden is not a static thing; it grows and changes. New plants should be tried, exotic as well as native. Our southern highways need beautification, and for this purpose some of our native trees, such as the flowering dogwood, red-buckeye, magnolias, red-bud, palms, and many others are unsurpassed. An appreciation of native plants and the natural gardens in our environment should be more generally developed that we may live better and happier lives.

Native

MAGNOLIA GRANDIFLORA (page 37)

WATER-PLANTAIN FAMILY (Alismaceae)

ARROW-HEADS (*Sagittaria*) Arrow-heads are recognized by their 3-petaled, white flowers borne in widely spaced whorls on the upper portions of the stem. The upper are usually staminate and the lower, pistillate. The name was suggested by the arrow-head-shaped leaves of certain species, but many species have differently shaped leaves. Lance-leaved arrow-head has lanceolate leaves up to 2′ long. *S. lorata* and others have no leaf blade but simply elongated leafstalks, while *S. stagnorum* has a few spade-like leaves more or less lobed at the lower corners. All inhabit marshy ground on margins of ponds, lakes, and streams and are therefore suitable for fountains and lily ponds.

white
1 pistilate
2 staminate

fl 3-leaf leaves
of 4 of 5

Sagittaria lancifolia 4. *S. lorata* 5. *S. stagnorum*

GRASS FAMILY (Gramineae or Poaceae)

SEA-OAT, SOUTHERN DUNE-GRASS (*Uniola paniculata*) A robust, light-green grass with showy, gracefully arched panicles of numerous large spikelets, growing in beach sand and forming dunes. Remarkable for its resistance to salt spray from ocean breezes. When stems are cut soon after the panicles are fully expanded, it makes handsome dried bouquets which will last for months and even years. Ranges from s. Fla. to Tex. and n. to s.e. Va. A closely related species, the inland sea-oat (*U. latifolia*), which grows along stream banks in the Piedmont and adjacent areas, may be grown as an ornamental and also makes decorative dried bouquets.

Uniola paniculata — *U. latifolia*

SEDGE FAMILY (Cyperaceae)

WHITE-TOP SEDGE (*Dichromena latifolia*) The small groups of inconspicuous flowers characteristic of the sedges are in this one surrounded by an involucre of white, pointed bracts 1½″-2″ long giving the inflorescence a composite-like appearance with pointed, drooping rays. It grows in colonies on low savannas among pitcher-plants and sundews and ranges from Fla. to Tex. and n. to Va. The showiness of this unique sedge is an outstanding feature of Coastal Plain vegetation during the early summer months. Two smaller and less conspicuous white-tops occur in the South.

[1]

white

Dichromena latifolia

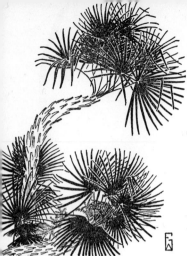

PALM FAMILY (Palmaceae)

SAW- or SCRUB-PALMETTO (*Serenoa repens*) The commonest palm seen in Fla. landscapes, savannas, hammocks, sand dunes, pinelands, and scrub. Variable in color and size. The stems are usually horizontal and will take root where in contact with the soil, but may be ascending or erect and attain a height of 20'. As a palm it is unique in that the stem is more or less branched. The name "saw-palmetto" alludes to the sharply serrate petioles. Flowers are white and fragrant, and the drupaceous fruits were an important food of the Indians. Fla. to La. and n. to S.C.

Serenoa repens - ivory-white

ARUM FAMILY (Araceae)

WHITE-ARUM (*Peltandra glauca*) The arums have a flower-stalk called a "spadix" which bears around its base, or throughout its whole length, groups of simple, inconspicuous flowers, the whole surrounded by a leaf-like structure called a "spathe." The spathe may be green or variously colored and usually forms the showy part of the inflorescence. In white-arum, the spathe is whitish and spreading. It commonly grows on borders of swamps in acid soil, associated with arrow-heads, pipeworts, yellow-eyed grasses, and other aquatics. Fla. to Ala. and n. to s.e. N.C. Green-arum (*P. virginica*) with a green spathe is widely distributed in e. N.A. Both may be grown in garden ponds.

Peltandra glauca white

GOLDEN-CLUB (*Orontium aquaticum*) In February and March when the red-maple is blooming and the springpeepers begin to sing, the "clubs" of the golden-club emerge from the brown-colored waters of shallow marshes and swamps of our Coastal Plain. The "clubs" are the flowering spikes or spadices which are surrounded by an inconspicuous rudimentary spathe and are covered to the tip with perfect yellow flowers. Its leaves are bluish-green and more or less iridescent. It may be completely submerged except for the spadix or more or less exposed. Fla. to La., n. to N.Y. and in some of the eastern inland states.

[2]

Orontium aquaticum-yellow

JACK-IN-THE-PULPIT (*Arisaema triphyllum*) In this curious plant the pointed extension of the spathe arches over the spadix suggesting the canopy of an old-fashioned pulpit under which "Jack [the spadix] preaches today." The spathe varies in color from pale-green to brown or purple-striped within. The fruits consist of a cluster of coral-red berries. A perennial growing from a bulbous corm with an acrid juice and minute needle-shaped crystals of calcium oxalate which together give a sharp, burning, prickly sensation when tasted. Related to the green-dragon (*A. Dracontium*), so-called in allusion to its long, attenuate spathe resembling the tail of a reptile. In moist, rich soil of various situations and widely distributed in e. N.A. Easily transplanted to flower gardens. Three other species occur in the South.

Arisaema triphyllum -green, red

YELLOW-EYED GRASS FAMILY (Xyridaceae)

ELLIOTT'S YELLOW-EYED GRASS (*Xyris Elliottii*) Yellow-eyed grasses are not grasses but rush-like, mostly perennials with basal leaves and a naked flowering stalk terminated by a scaly flowering head. The 3 delicate petals are usually yellow, rarely white. About 18 species occur in the South. The one selected for the illustration here is named for the noted S.C. botanist, Stephen Elliott (1771-1830). The majority of yellow-eyed grasses grow in moist to wet situations chiefly in the Coastal Plain. Commonly seen in roadside ditches throughout the summer months. Used for fountain borders.

PIPEWORT FAMILY (Eriocaulaceae)

HAT-PINS, TEN-RIBBED PIPEWORT (*Eriocaulon decangulare*) The hat-pins or pipeworts are among our most conspicuous and interesting roadside plants throughout the Coastal Plain during the summer months, associated with yellow-eyed grasses, sundews, rushes, and sedges. The small white flowers are crowded in a button-like head at the end of a naked stem 1'-3' tall. They grow in depressions in savannas, edges of swamps, and in roadside ditches. This is the most common species which ranges from Fla. to Tex. and n. to N.J. Three other pipeworts occur in the South. Related to pipeworts are shoe-buttons (*Syngonathus flavidus*), with short, hairy stems and straw-colored heads, and hairy-pipeworts (*Lachnocaulon* spp.), with gray heads. All are excellent for dried bouquets.

Xyris Elliottii-yellow- pod-dried arr.

[3]

dish gardens
dried arr.

white

Eriocaulon decangulare-Lachnocaulon.

blue 1 | pink 1 | blue
Commelina angustifolia 2 *Cuthbertia graminea*

stamen X

blue - *Tradescantia reflexa*

violet-blue

Pontederia cordata

SPIDERWORT FAMILY (Commelinaceae)

ROSELING (*Cuthbertia graminea*) This low, delicate pink-flowered, deep-rooted perennial is closely related to spiderworts. The 3-petaled flowers with conspicuously hairy filaments within appear in the spring among its grass-like leaves. In sandy soil, Coastal Plain, Fla. to s.e. Va.

NARROW-LEAVED DAY-FLOWER (*Commelina angustifolia*) Few flowers surpass the delicacy of the cerulean blue of the narrow-leaved day-flower which blooms in the early morning but whose petals soon wither with the advancement of day. As in many day-flowers the upper 2 petals are large and showy while the lower is small and inconspicuous. The 6 stamens also differ in shape, color, and function. A deep-rooted perennial of mostly sandy soil of the Coastal Plain and adjacent areas. Fla. to Tex. and Va.

SPIDERWORT or BLUE-JACKET (*Tradescantia reflexa*) Spiderworts are recognized by their blue or violet (rarely white or red), delicate 3-petaled flowers with conspicuously hairy filaments and long, gradually tapering leaves. About 9 species occur in the southern states. The one illustrated is distinguished by its deep-blue petals, sepals with tufts of hairs at the tips and smooth stems. It grows in various situations but mostly in sandy soil and blooms in late spring and summer. It ranges from Fla. to Tex. and N.C. This and other species are often transplanted to gardens of country homes where they thrive with little care.

PICKEREL-WEED FAMILY (Pontederiaceae)

PICKEREL-WEED, WAMPEE (*Pontederia cordata*) The bright bluish-lavender spikes of the native southern "water-hyacinth," as it is sometimes called, present a pageant of bloom from June to September, or all year in the southern portions of our range. The ephemeral flowers have one orchid-like lip, like that of the true water-hyacinth, but darker in color. Each flower lasts but a day as new ones open progressively below the growing tip of the spike. Ever present on muddy shores of ponds, streams, swamps, and lakes, associated with arrow-heads, arums, golden-clubs, and other aquatics. Although most frequent in the Coastal Plain, it occurs in various provinces throughout e. U.S. and Can.

[4]

PINEAPPLE FAMILY (Bromeliaceae)

This family is noted for its epiphytes, plants growing on other plants but deriving no nourishment from them, generally known as air plants or bromeliads.

SPANISH-MOSS (*Tillandsia usneoides*) (color page facing 145) The long, gray streamers of this remarkable plant are a distinctive feature of many southern landscapes. It is not a moss but a flowering plant related to other air plants and the pineapple, and is not a parasite as is sometimes thought. It bears numerous small, greenish-yellow flowers which form interesting seeds with hairy sails similar to those of many composites. Especially abundant on live-oaks and pines along estuaries and borders of rivers and swamps. Fla. to Tex. and Va.; also W.I.

BALL- or BUNCH-MOSS (*Tillandsia recurvata*) This is related to Spanish-moss but has closely overlapping leaf-bases and long-attenuate, flexuous blades with curved tips, covered with a grayish fuzz. A long, wiry stalk bears at its summit a small group of inconspicuous violet flowers. Common in hammocks, swamps, and low pinelands and often on wires and posts. Fla. to Tex. and tropical America.

SLENDER-LEAVED TILLANDSIA (*Tillandsia tenuifolia*) One of the smallest and most delicate of air plants inhabiting trees of low grounds. Its slender green leaves which are widened at base often turn purplish-red during the flowering period. These together with its reddish flowering stalks and small deep violet blossoms contribute to its distinctive ornamental value. Fla to Ga.; tropical America. This and other species are often sold as "orchids."

WILD PINEAPPLE (*Tillandsia fasciculata*) (color page facing 145) This is the largest and showiest of our air plants. The large leaves are a light gray-green most of the year but change to a silvery purplish tint during the flowering period. In early spring the stout, bracted, usually branching, flowering stems appear, colored a brilliant vermilion, terminated by the long-lasting red-bracted spikes of inconspicuous violet flowers. It is an impressive sight to see the huge clumps of these plants in spring in their height of color on the branches of giant cypress trees in hammocks and swamps. Fla. and tropical America.

[5]

yellowish   blue

Tillandsia usneoides – *T recurvata*

Tillandsia tenuifolia – dark red b.   flower violet

Tillandsia fasciculata   bracts red & yellow   flower violet

LILY FAMILY (Liliaceae)

WHITE-FEATHERLING (*Tofieldia glabra*) The delicate, white racemic spikes of this small-flowered liliaceous perennial may be seen on open pinelands during summer and early autumn. Similar and related to the false-asphodels, or wood-featherlings, of wider and more northern distribution, but distinguished from these by its smooth stems and leaves. It ranges in the Coastal Plain from Ga. to N.C. Keeps well when cut and may be cultivated in acid soil. A closely related plant, the rush-featherling (*Pleea tenuifolia*) blooms in late August and September, associated with white- and yellow-fringed orchids. It is recognized by its rush-like appearance and its white, star-shaped flowers borne in a bracted spike-like raceme. Ga. to N.C.

Tofieldia glabra-white

FAIRY- or ANGEL'S-WAND (*Chamaelirium luteum*) In this attractive plant the sexes are separated so that plants are either male and furnish pollen or female and bear the fruit. Flowers are borne in spike-like racemes and those of the male are longer and much showier than the female. When cultivated in rich soil, the former become long and nodding as in the lizard's-tail. It grows in moist, rich wood soil from the Coastal Plain to the mountains and is widely distributed in e. N.A.

Chamaelirium luteum-white

FEATHERBELLS, FEATHER-FLEECE (*Stenanthium gramineum*) This is well named for when in full bloom, its large panicles resemble large, white feathers. The small, white flowers with pointed sepals and petals may be male, female, or bisexual in the same inflorescence. The stem is from 3′-6′ tall and the leaves, which are mainly basal, are long and narrow. It grows in rich woods, thickets, and meadows of upland districts. A related robust form with thick, tall, and leafy stems and a congested inflorescence is by some considered only a variety of the above and by others worthy of specific rank. This is the form which is often cultivated. Both are widely distributed from Pa. southward but the robust form seems to be more northern in its range. Both bloom from June to September.

[6]

white

Stenanthium robustum - S. gramineum

OSCEOLA'S-PLUME, CROW-POISON (*Zigadenus densus*) The thick, candle-like racemes of Osceola's-plume are the most conspicuous of our Coastal Plain flowers during April and May. The relatively large white flowers are numerous at the summit of a stem 20"-40" tall. The long strap-shaped leaves are mainly basal. Moist pinelands, borders of pocosins and swamps, Fla. to Miss., and s.e. Va.

SOUTHERN-CAMAS (*Zigadenus glaberrimus*) The petals of the white flowers borne in a broad, loosely pyramidal panicle have no stalks and bear 2 greenish glands at base. Stems are 2'-3' tall with elongate, strap-shaped leaves which are mainly basal. Resembles bunch-flowers (*Melanthium*) but distinguished by its sessile petals and smooth foliage. Moist pinelands, Coastal Plain, Fla. to La. and Va.

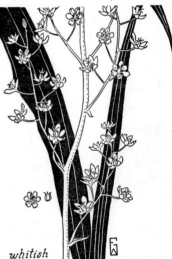

white

Zigadenus densus-Z.glaberrimus

SOUTHERN BUNCH-FLOWER (*Melanthium dispersum*) Four species of bunch-flowers occur in e. U.S. Of these the southern bunch-flower has the most limited distribution, ranging from Fla. to Miss. and N. C. A coarse perennial with strap-shaped leaves which are mainly basal. The sepals and petals are greenish-white, elliptic, stalked, and bear 2 glands at base. The panicle is few-flowered and about as broad as long. Distinguished from other bunch-flowers by its elliptic perianth parts and diffuse panicle. This as well as other bunch-flowers should be cultivated more than they are.

whitish

Melanthium dispersum

STRIPED GARLIC (*Allium Cuthbertii*) Few who are not botanists are aware that onions and garlics are related to the lilies and their relatives. Several of the wild ones have beautiful flowers which vary in color from white to pink or purple. Some, such as the common field-garlic, an obnoxious lawn weed, bear mostly bulblets in place of flowers. However, striped-garlic does not bear bulblets in the inflorescence which consists of an umbel of white flowers. Sepals and petals are similar with gradually pointed blades, and the stamens are alternately large and small. It grows in sandy woods of the outer Coastal Plain from Fla. to Ala. and N.C. The related *A. tricoccum,* called ramp, is highly prized for eating by our mountain people.

Allium Cuthbertii

Lilium Catesbaei - *red-speckled*

SOUTHERN RED-LILY, CATESBY'S-LILY (*Lilium Catesbaei*) (color page facing 64) This lily which was named for one of the earliest southern naturalists, Mark Catesby (1679-1749), is the only red lily with erect flowers in the Coastal Plain. Sepals and petals are narrowed below into distinct claws and taper gradually above into attenuate tips which are outwardly curved. The stem bears alternate leaves which are reduced in size upward. It grows from bulbs with slender scales. Moist, open pinelands and borders of swamps, Fla. to La. and N.C. Reported difficult to transplant. A form with blunt, oblong lower leaves and more erect, less outwardly curved perianth tips is var. *Longii*. Ala. to s.e. Va.

Lilium Michauxii - *orange spotted*

CAROLINA-LILY (*Lilium Michauxii*) This lily was discovered by the noted French botanist André Michaux (1746-1802) who named it *L. caroliniana,* but since this name was found to be untenable it was later named for its discoverer. One to 3 flowers are borne at the branching summit of a stem up to 4' tall. The whorled leaves are broadened outward and blunt or abruptly short-tipped. Pine and oak woods, especially on mountain slopes and summits, n. Fla. to La. and Va. This is often associated with Gray's- or roanlily (*L. Grayi*), named for its discoverer Asa Gray (1810-1888). Both are often cultivated.

Lilium superbum - *orange spotted*

TURK'S-CAP, LILY-ROYAL (*Lilium superbum*) One of the tallest and most spectacular of the red lilies with panicles of 3 to many nodding, showy flowers with strongly reflexed perianths, borne at the top of a stem up to 8' tall. The perianth parts are golden at base spotted with maroon. The long, narrow leaves occur in a series of whorls except a few upper ones which are alternate. Margins of blades are smooth or with low, rounded papillae. Various upland situations but most gorgeous in rich mountain valleys. Ga. to Ala. n. to N.H. Often cultivated.

[8]

YELLOW ADDER'S-TONGUE, TROUT-LILY (*Erythronium americanum*) A true harbinger of spring, the yellow adder's-tongue, or dog-toothed violet as this lily has been misleadingly called, is one of the first perennials to bloom in our area. It occurs typically on rocky, north-facing stream bluffs in the Piedmont and mountains. At bases of bluffs where the soil is deep from periodic deposit of soil by flooding, the flowers may become twice as large as the average. The brown spots on the two leaves suggested to John Burroughs the names "fawn" or "trout-lily." Widely distributed in e. N.A. ranging as far south as n. Fla. It is quite variable so that some of our southern forms may eventually be recognized as distinct species.

Erythronium americanum yellow purple lines

EASTERN-CAMASS, WILD-HYACINTH (*Camassia scilloides*) The eastern-camass bears showy racemes of relatively large, slightly irregular, pale-blue flowers on a naked stem, or scape. It grows from bulbs 1"-2" long. Leaves are strongly keeled, long, and usually narrow but may be broad and sometimes variegated. The fruit is a triangular-globose capsule. Grows in uplands, open woods, meadows, and low fields from Ga. to Tex. and n. to several n.e. states. It is easily cultivated in suitable situations.

Camassia scilloides -white

TAWNY or ORANGE DAY-LILY (*Hemerocallis fulva*) After being introduced from Eurasia, this lily soon escaped from gardens and has now spread and established itself along roadsides, borders of fields and woods, and various other situations throughout e. U.S. As it is self-sterile and rarely sets seed, its spread depends upon the distribution of its fleshy roots which are produced in great abundance and are very tenacious of life. While many beautiful varieties of day-lilies are now used in gardens, some gardeners prefer this naturalized species which, once established, is difficult to eradicate. A double variety native of Japan (var. *Kwanso*) expands over a longer period.

[9]

Hemerocallis fulva - tawny-orange

Clintonia umbellulata-white

SPECKLED WOOD-LILY, GLOBE-AMORETTE (*Clintonia umbellulata*) The umbel of white flowers with spreading petals and sepals speckled with green and purple, borne at the end of a naked stalk, and a few broad basal leaves are the marks of recognition of this attractive native flower. It grows in rich, moist wood soil in the mountainous districts descending along stream bluffs to the foothills. Since it is not common, it is always thrilling to discover this lily. It ranges from Ga. to N.Y. The northern-clintonia (*C. borealis*), with yellow flowers, occurs on some of the higher mountains.

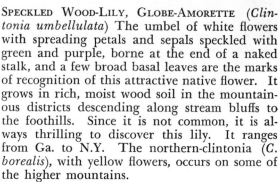

cream

Yucca aloifolia   Yucca filamentosa   ×¾

SPANISH-BAYONET (*Yucca aloifolia*) Because of their adaptation to dry environments, Yuccas add picturesqueness to vegetation. Spanish-bayonet is typically coastal, growing in beach sand and among dunes but thrives when planted inland in better soil. The stems are sparingly branched, up to 8′ tall, and covered with hanging dead leaves except in the upper portion. **Gorgeous** panicles of creamy-white flowers are produced in early summer and again in late summer. The principal mode of reproduction and spread is by pieces of stems drifting ashore. The dagger-like leaves are dangerous to come in contact with. Fla. to La. and N.C. Spanish-dagger (*Y. gloriosa*) is a similar species except for its short stems and long flowering stalks. It grows naturally further inland. Two related species (*Y. filamentosa* and *Y. Smalliana*), with flaccid leaves, fibrillose on margins, and short stems, are widely distributed in the South.

Medeola virginiana-white

INDIAN CUCUMBER-ROOT (*Medeola virginiana*) The 2 whorls of leaves and the pale greenish-yellow flowers borne on recurved pedicels below the upper whorl are the outstanding characteristics of this herbaceous perennial. It is the only species of this genus. It grows from white, horizontal tubers with a taste of cucumber, hence its common name. Rich soil of the upper Piedmont and mountains, Fla. to La. and n. to n.e. states and Can. An attractive plant which should be in every wild flower garden.

[10]

IRIS FAMILY (Iridaceae)

BARTRAM'S-IXIA (*Ixia coelestina*) This beautiful southern flower was discovered and named by William Bartram (1739-1823) in his famous *Travels* in which he mentions the "azure fields of coerulean Ixias." The violet, rarely white, petals with white "eye" open at sun-up but soon fade away. Locally they are called "violets." Flatwoods of n. Fla. and s. Ga.

Ixia coelestina - mauvette

CELESTIAL-LILY (*Nemastylis floridana*) This exquisite, fragile "lily" with delicate, violet-blue, ephemeral deliquescent flowers differs from the true Ixias in blooming in the fall from mid-afternoon to sundown while the latter bloom in the forenoon of spring days. It grows from a bulb with a sparingly branched stalk with narrow, folded leaves. Very restricted in distribution, confined to swamps, marshes, and flatwoods of n.e. Fla. Easily grown from seed. *N. geminiflora* (*N. acuta*) has a wider distribution, growing in prairies, calcareous glades, and rich woods, La. to Tex. to w. Tenn., Mo. and s.e. Kans. It has larger flowers than the above and blooms from forenoon to mid-afternoon.

Nemastylis floridana - mauvette

HERBERTIA (*Herbertia Drummondiana*) Because of its rarity and limited distribution this beautiful flower has escaped a common name. One or 2 flowers are borne from a spathe at the summit of a leafy stem. The sepals are pale-violet with a darker violet band outlining the white base which is violet spotted. Petals are light-violet in the upper part becoming dark-violet below and white spotted at base. Prairies and marshes, Coastal Plain, La. to Tex.

BLUE-EYED GRASSES (*Sisyrinchium*) Because of their grass-like habit and small flowers it may seem strange that the blue-eyed grasses are relatives of iris. They are mostly American perennials with fibrous roots, narrow leaves, and slender stems. About 16 species have been reported from the southern states. Of these, *S. atlanticum* is the most beautiful. Sandy soil, near the Coast. Fla. n. to e. Can.

Lavender   *blue*

Herbertia Drummondiana

Sisyrinchium atlanticum

LOUISIANA IRISES

The Irises of the lower Mississippi Delta of Louisiana are perhaps the most remarkable in all the world. Although Small lists 80 species, many are thought to be hybrids or of hybrid origin. They are now being transplanted for preservation and study by Miss Caroline Dorman of Saline, La., author of *Wild Flowers of Louisiana*. In the color illustration opposite, a few of the Louisiana Irises are represented. At the top is the naturalized European *Iris Pseudoacorus* which also grows in this area. On the left of this is the red iris (*I. fulva*) and in the right foreground, *I. giganticaerulia*. The remaining 2 are unidentifiable forms probably of hybrid origin.

WILD POPPY MALLOW (*Callirrhoë Papaver*)   Mallow Family

Poppy-mallow is an herbaceous plant with procumbent stems and 3- to 5-lobed, or parted leaves. Flowers have red or reddish-purple petals and no bracts below the calyx. A showy, attractive native blooming in spring and summer in sandy woods, Coastal Plain, Fla. to Tex. and Ga.

NATIVE RED HIBISCUS (*Hibiscus coccineus*)   Mallow Family

The most gorgeous of our native flowers with its spreading, crimson corollas up to 6" across. A perennial up to 10' tall with palmately divided leaves. In swamps and other low areas near the coast from Fla. to Ala. and Ga. Cultivated in drier soils outside of its natural range as far n. as N.C.

WATER HYACINTH (*Eichornia crassipes*)   Pickerel-Weed Family

This showy aquatic has beautifully formed lilac flowers, blotched with blue and yellow. The frail blossoms remain open but a day and soon wilt when picked. There is, however, a continual sequence of flowers from early spring till late fall. Supposedly introduced from S.A., this attractive aquatic multiplies so rapidly that it has become a troublesome "weed" in many lakes and streams in the southern parts of its range. Fla. to Tex., n. to Mo. and Va.

PITCHER PLANTS (*Sarracenia*)   Pitcher-Plant Family

Pitcher plants are remarkable for their hollow, pitcher- or trumpet-shaped leaves which are often provided with a canopy over the top. The leaves are usually more or less colored in light-green or yellows and lined with red or purplish. One flower is borne at the summit of a naked stalk. The most remarkable feature of the flower is its large umbrella-shaped stigma. At the left in the illustration is the common yellow pitcher-plant (*Sarracenia flava*) of the Coastal Plain from Fla. to Ala., n. to Va.; at the right is the more ornamental *S. Drummondii*, restricted to sandy bogs of n. Fla. to Miss. and Ala.

Louisiana Irises

Native Red Hibiscus and Wild Poppy Mallow

Water Hyacinth

Pitcher Plants

SPRING WILD FLOWERS OF THE SOUTHERN COASTAL PLAIN
Collected at Highlands Hammock, Sebring, Florida, in late March

SPRING WILD FLOWERS OF THE SOUTHERN COASTAL PLAIN

1. Wild Poinsettia
2. Blue Flag, Iris
3. Rattle Box (escape)
4. Osceola's Plume
5. Amsonia
6. False Dragonhead
7. Grass Pink
8. Vanilla leaf, Deer Tongue
9. Florida Dandelion
10. Sensitive Brier
11. Diapedium
12. Thimbles
13. Indian Shot, Canna
14. Thistle
15. Primrose Willow
16. Colic Root
17. Butterfly Weed, Milkweed
18. Spider Lily
19. Beard Tongue
20. Button Pipewort
21. Southern Elder
22. Rattle Box (escape)
23. Pickerel Weed
24. Arrow Head
25. Rushweed

26. White Top Sedge
27. Red Sage
28. Tar Flower
29. Sunflower
30. False Foxglove
31. Dwarf Clematis
32. Yellow-Eyed Grass, Hardhead
33. Thimbles, Orange Milkwort
34. Pink Milkwort
35. Grass Pink Orchid
36. Blue-Eyed Grass
37. Trumpet Vine
38. Sand Pea
39. Lizard's Tail
40. Star Grass
41. White Topped Aster
42. Blue Dew Flower
43. Butterfly Pea
44. Tread Softly, Spurge Nettle
45. Beggar's Ticks
46. Black Eyed Susan
47. *Calophanes oblongifolia*
48. Passion Flower
49. Vetch
50. White Violet

51. Morning Glory
52. *Rotantha floridana*
53. Rose Pogonia
54. Skull Cap
55. *Ammopursus Ohlingeri*
56. Chapman's Pea
57. Rabbit Tobacco
58. Meadow Beauty
59. Blue Lupine
60. Marsh Pink
61. Horse Nettle (fruit)
62. Paw-paw
63. Blue Sage
64. Wild Coreopsis
65. Blue Spiderwort
66. Bishop's Weed
67. Tall Milkwort
68. Hairy Ruellia
69. Milkweed
70. Yellow Aster
71. Goldenrod
72. Cherokee Bean
73. Prickly Pear Cactus
74. Blue Hearts

[17]

Few flowers are in general better known and used more for ornamental purposes throughout the world than Irises or flags. The name Iris, which comes from a Greek word meaning rainbow, is very appropriate as nearly all colors of the spectrum are exhibited by this genus. Even a single species, or variety, may show a remarkable variety of colors in each flower. Flowers are borne in special bracts, or spathes, and in some, such as the dwarf irises, each flowering stem bears only one flower. In the majority of species, however, several flowers are borne in succession on the stem as well as from each spathe. In the latter the bracts are called "heads." The parts of the flower are united into a longer or shorter tube with the seed pod at the base. All parts of the flower, except the stamens which are hidden under the style lobes, contribute to the showiness of the flower. The outermost set, composed of the sepals and often called "falls," is the showiest part. The sepals are not only larger than the other parts, but are often ruffled in various ways, differ in ground color from the other parts, and are variously dotted, streaked, or splotched in different colors and crested with hairs or ridges. The petals, which alternate with the sepals, are usually smaller, and in one of our native species, *I. tripetala,* are so reduced as to be hardly visible. The flattened style branches are also colored and split or crested at the tip which extends beyond the stigma surface beneath. Many of our cultivated Irises are of exotic origin and consist mostly of cultural forms and hybrids. The common yellow-iris (*I. Pseudoacorus*) of the e. U.S. is an exotic which has become naturalized in this country. In number of forms and variety of colors, the southern native Irises are the most remarkable in all the world. In his *Manual of the Southeastern Flora* (1933), Small lists 95 species. The majority of these occur in the Mississippi Delta of La., a veritable natural iris garden discovered by Dr. Small in 1925. Here the colors range from blues through pinks, fulvous, yellows, reds to velvet-purplish. However, many of these "species" show such a wide range of colors and forms that it is questionable if they should be classified as true species. In fact, several of them have been found to be of hybrid origin. Nevertheless, here is a vast amount of beautiful material of the highest ornamental value which can be used for culture as well as for the production of new forms through selection and artificial hybridization.

DWARF-IRISES *(Iris)* Two species of dwarf-irises occur in the South; in one, *Iris verna,* the leaves are narrow and erect, the sepals light-blue with a papillose light-yellow crest. It is delicately fragrant. Grows in open, acid woods and dry pinelands in various sections but most abundant in the Coastal Plain. Ga. to Miss., Ky. and Pa. In the crested dwarf-iris, the leaves are broader, more curved and spreading than in the above. The sepals have a beaded crest bordered with violet. The ground-color is violet-purple. Rich, moist woods and stream bluffs. Ga. to Ark. and Md. White forms occur occasionally.

Iris cristata-blueviolet-I.verna - violet

SHORT-PETALED IRIS *(Iris tripetala)* A beautiful species easily recognized by its apparently 3-parted, instead of 6, perianth parts—and because of its short petals. Sepals deep violet, yellow blotched at base. Low pinelands and borders of swamps, Coastal Plain, n. Fla. to N.C.

SOUTHERN BLUE-FLAG *(Iris virginica)* The most frequent and widespread of our tall flags. Often seen in roadside ditches and borders of marshes and swamps. Color of perianth varies from dark-violet to pale-pink, rarely white. Coastal Plain, Fla. to Tex. and Va. Similar to *I. versicolor* (color page facing 32), occasional in the mountains. *I. hexagona,* as the name suggests, is a species with 6-angled pods. The flower stalks are zigzag but erect; leaves, half-ovate. The flower parts are a rich violet but the sepals have a yel-low crest, whitish at base. A white form is known. Swamps, ditches, and stream banks. Coastal Plain, n. Fla. to Ga.

Iris tripetala-violet-I.virginica-lavender

FLORIDA-IRIS *(Iris savannarum)* This species also has a 6-angled pod and is one of our most beau-tiful species. Sepals have a compound crest with radially spreading branches of violet or bluish-violet ground-color, whitish or white-flecked at base. Petals are bluish to violet and style lobes, a greenish ground-color, flushed and streaked red-violet. In some places covers extensive areas on prairies and savannas. Peninsular Fla.

[19]

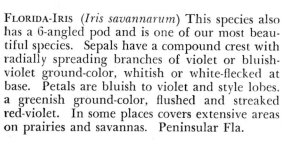

Iris savannarum-variable-I.hexagona-violet

BLOODWORT FAMILY (Haemodoraceae)

GOLD- or GOLDEN-CREST (*Lophiola aurea*) This attractive, perennial herb with slender rootstocks has a whitened stem with soft, matted, wooly hair. The small, yellow flowers, hairy within, are borne in forked cymes at the top of the stem. The narrow, grass-like leaves are hairy also. It is very conspicuous especially in the autumn sunshine in moist pinelands and acid bogs of the Coastal Plain from Fla. to Miss. and s.e. N.C.

Lophiola aurea-yellow, rust

CANNA FAMILY (Cannaceae)

GOLDEN-CANNA (*Canna flaccida*) This is our only native canna which is easily distinguished by its yellow perianth from the naturalized red-flowered Indian-canna (*Canna indica*). The showy spikes of several irregular flowers, 3″-4″ long, rise to a height of 2′-4′. One peculiarity of the flowers is the petal-like filaments which are united in a frilly tube around the one large upright stamen. The 6 long, thin petals droop lazily producing a ragged effect. It is frequent in marshes and swamps of the Coastal Plain from Fla. to S.C. Suitable for borders of ponds and in roadside ditches.

Canna flaccida - yellow

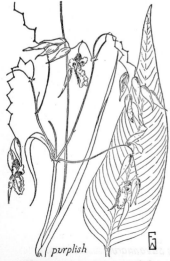

ARROWROOT FAMILY (Marantaceae)

THALIA (*Thalia geniculata*) The Thalias are the only native representatives of this interesting family which is closely related to the orchids. They are stout, scapose, perennial herbs, often white-powdered at least above, with broadly-lanceolate leaves and numerous purplish flowers in arching panicles. Being aquatics, they inhabit borders of ponds, rivers, and lakes. The natural range of the above species is from Fla. to the W.I., but when cultivated it is hardy as far north as N.C. The other species (*T. dealbata*), occurring in the South, ranges from Fla. to Tex., Mo. and S.C.

[20]

purplish

Thalia geniculata purplish

Orchids are the elite of the plant world. In their specialization for insect pollination, they have attained a superior distinctness in flower form and delicacy of shades of color. The extreme interrelationship between certain orchids and their pollinators is difficult to comprehend. In some cases, only one sex (the male) of one kind of insect can pollinate the flowers. In their adaptation to habitat, they are also in general highly specialized. Many are limited in their distribution and, where found, do not usually occur in great abundance. Still the orchids are widely distributed in the world from the tropics to the subarctic and compose one of the largest of plant families. It has been estimated that over 15,000 species of orchids exist in the world. About 85 per cent occur in the tropics, but even these do not always occur in warm climates. What they all seem to require, however, is a more or less humid atmosphere. As to habitat, they may be classed into terrestrial and epiphytic. The former are more widely distributed and are the only ones which extend into the temperate and subarctic regions, while the latter are restricted to the tropics and subtropics. It is among the tropical, epiphytic orchids that the most gorgeous flowers are found. The majority of orchids obtain their nourishment as other green plants do, but some live saprophytically on the leaf mold after the fashion of mushrooms. They reproduce by seed and to some extent vegetatively. The seeds of orchids are the smallest of all plant seeds and are produced in enormous numbers. It has been estimated that the number of seeds in one capsule of many orchids exceeds a million. Natural germination of most orchids requires the association of a mold. However, it has been discovered that by adding the sugar glucose to a nutrient medium, they germinate readily without the mold. The orchid flower consists of 3 sepals which are usually similar in shape and color and 3 petals, 2 of which are not much different from the sepals and one, known as the "lip," which generally differs considerably from the others both in size and shape. It ranges from a pouch as in the lady-slippers to a fringe of hairs as in the fringed-orchids. Most lips are provided at base with a nectar spur. The lip is situated on the upper side of the flowers, but because of the twisting of the pedicel, it becomes displaced to the lower side. In Calopogons or grass-pinks, it retains its original upper position. The stamens are reduced to one or 2 and are joined with the stigma forming a structure known as a "column." Many of the southern native orchids are difficult to culture even in the vicinity where they grow naturally. This is largely due to their sensitivity to the soil conditions, especially to its acidity and moisture. Cultural methods for our native orchids have been ably described by Dr. E. T. Wherry of the University of Pennsylvania in Donovan S. Correll's book, *Orchids of North America* (1950).

CAPE-SABLE ORCHID, DINGY-FLOWERED ONCIDIUM (*Oncidium luridum*) This is the largest and most conspicuous of our tropical, epiphytic orchids, growing especially on bald-cypress in hammocks of s. Fla. Plants are large and coarse with a solitary leaf at the end of each pseudobulb (short, thickened, bulb-like stems). The inflorescence consists of a long spray-like, loose panicle. The perianth parts are broad and conspicuously stalked and mainly orange, mottled with brown. S. Fla. to W.I., Mex., C.A. and S.A.

yellow brown spots

Oncidium luridum "Cape Sable Orchid"

FLORIDA-ONCIDIUM (*Oncidium floridanum*) This happens to have a very appropriate specific name, for it is the only species of *Oncidium* restricted to Fla. Plants are slender, often growing in clumps, with long, pointed, plicate leaves. The inflorescence is a diffuse panicle with yellow flowers, mottled with brown, pointed and sessile. It flowers sporadically throughout the year but mostly from May to August, and in November and December. While it is commonly epiphytic it also grows on soil and on decaying logs in low pinelands and hammocks. It has been recorded for Collier, Dade, and Monroe counties.

brown spots on

Oncidium floridanum — yellow

CLAM-SHELL ORCHID (*Epidendrum cochleatum*) A distinctive species by its shell-like, purple lip and narrow, greenish-white, or yellow perianth with purplish spots at base. The Fla. specimens are unique in having 3 stamens instead of 1 (var. *triandrium*). Epiphytic in hammocks and cypress swamps, Fla.; also in tropical Americas.

FALSE-FLOWERED POLYSTACHYA (*Polystachya luteola*) Small, curiously-shaped, yellow flowers in many racemes characterize this epiphytic orchid. The hood-like shape of the flowers is due to the upwardly-folded sepals attached to the column. Widespread and rather frequent in Fla.; tropical America and Old World tropics.

[22]

Polystachya luteola-Epidendrum cochleatum.

GREEN-FLY ORCHID (*Epidendrum conopseum*)
This is the most frequent and widely distributed
epiphytic orchid in the southeastern states, as well
as the hardiest, for it extends farther north than
any other epiphytic orchid in N.A. Flowers are
relatively small in racemes, grayish-green, occasionally tinged with purple, with a delicate fragrance. It is found on various species of trees
associated with the gray-polypody fern and Spanish-moss, ranging from s.e. N.C. to s.c. Fla. and
w. to La.; in Mex. represented by var. *mexicana*.
Sometimes associated with this in Fla. is the
small, white, spectral palm-polly (*Polyrrhiza
Lindenii*).

BROWN-EPIDENDRUM (*Epidendrum anceps*) This
epiphytic orchid bears a group of small flowers
at the end of a leafy stem without pseudobulbs.
Flowers are fleshy, light greenish-brown to dull-
red or tawny-yellow. Frequent and widespread
in s. Fla.; Mex., C.A., W.I. and n. S.A. *E. nocturnum*, called "night-smelling epidendrum" because of its fragrance being especially noticeable
at night, also has a leafy stem, but flowers are
few, large and showy with long, somewhat
twisted, greenish-white sepals and petals. The
white lip is 3-lobed, the middle attenuate to a
slender point. Distribution, similar to the above.

YELLOW LADY'S-SLIPPER (*Cypripedium Calceolus*)
The Cypripediums are orchids with a characteristic sac-like lip which, owing to their fancied resemblance to a shoe, have been given the names
lady's-slippers and moccasin-flowers among many
others. The yellow lady's-slipper has a leafy
stem and the lip is not split on the upper side
as in the pink lady's-slipper (*C. acaule*) (color
page facing 32). It is widely distributed in N.A.
and is highly variable. The larger, more typical
form which is more frequent in the South from
Ala. and Ga. northward is var. *pubescens*. A
smaller form of northern affinities which occurs
in the s. Appalachian Mountains is often desig-
nated as var. *parviflorum*. Both are becoming
rare and in danger of being exterminated.

[23]

greenish
Epidendrum conopseum

cream   2 greenish or purplish
1 *Epidendrum* anceps 2 *E. nocturnum*

Cypripedium Calceolus v.*parviflorum*

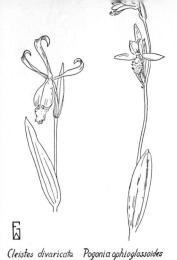

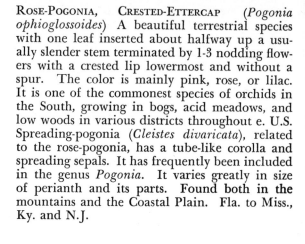

ROSE-POGONIA, CRESTED-ETTERCAP (*Pogonia ophioglossoides*) A beautiful terrestrial species with one leaf inserted about halfway up a usually slender stem terminated by 1-3 nodding flowers with a crested lip lowermost and without a spur. The color is mainly pink, rose, or lilac. It is one of the commonest species of orchids in the South, growing in bogs, acid meadows, and low woods in various districts throughout e. U.S. Spreading-pogonia (*Cleistes divaricata*), related to the rose-pogonia, has a tube-like corolla and spreading sepals. It has frequently been included in the genus *Pogonia*. It varies greatly in size of perianth and its parts. Found both in the mountains and the Coastal Plain. Fla. to Miss., Ky. and N.J.

Cleistes divaricata   Pogonia ophioglossoides

GRASS-PINKS, ROSE-WINGS (*Calopogon pulchellus*) Few flowers have been so extravagantly named, for *Calopogon* means "beautiful beard" in allusion to its mixed, purplish, orange, cream, and reddish hairs on its uppermost lip, and *pulchellus* means the "little beauty." It grows in acid soil in various situations in e. N.A. Many-flowered grass-pink (*Calopogon multiflorus*) resembles superficially the above but has more and smaller flowers. Grows in damp, sandy pinelands, and flatwoods. Associated with saw-palmetto, Fla. to Ala., Miss. and Ga.

pinkish

Calopogon pulchellus-C. pulchellus var multiflorus

MAUVE SLEEKWORT, LARGE TWAYBLADE (*Liparis liliifolia*) Although fairly frequent in various provinces of its range, it is always thrilling to find this low, terrestrial orchid with its mauve-purple, wedge-shaped lip tinged with green. It grows in soil or in tufts of mosses on flood plains and river bluffs, blooming from May to July. The two leaves at its base are glossy and keeled beneath. It reproduces vegetatively by pseudo-bulbs borne below the leaves. Found in various districts but mostly in the Piedmont and mountains, Ga. to Ala. and northward.

[24]

Liparis liliifolia -greenish-brown lip

LADIES'-TRESSES, SPIRAL-ORCHIDS (*Spiranthes*)
The generic name of these orchids was suggested by the usual spiral arrangement of the small flowers in the spikes. The flowers are mostly white but may be yellow or greenish. There are many species some of which are difficult to distinguish. The most frequent and generally distributed in the South is *S. cernua,* called nodding ladies'-tresses which blooms in late summer and autumn. A robust form of this species, which is usually fragrant, has been called var. *odorata.* This species is widely distributed in N.A. and is, therefore, as is to be expected, widely variable. Two tropical species occur in peninsular Fla. One of these is *S. cranichoides* with pink flowers. It occurs in rich soil, decaying leaves and humus, or on rotten wood in hammocks and moist woodlands. The other is *S. orchioides,* commonly called leafless beaked-orchid. Flowers are showy, varying from almost white to red or crimson. It occurs in dry waste places or along roadsides and in pinelands.

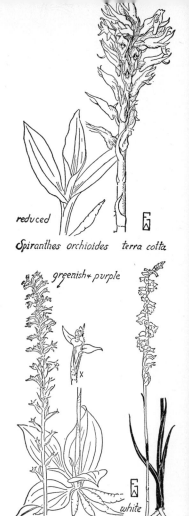

reduced

Spiranthes orchioides terra cotta

greenish + purple

x

white

Spiranthes cranichoides · S. cernua

CORAL-ROOTS, CORAL-ORCHIDS (*Corallorhiza*)
Some orchids do not have chlorophyll with which to make their own food, so they live somewhat like mushrooms on leaf mold. They are generally pink or brownish in color. Flowers are small in a terminal spike. Most of them bloom in the fall, but *C. Wisteriana* blooms in the spring. It extends from Pa. to c. Fla. Another saprophytic orchid in the South is crested coralroot (*Hexalectris spicata*) which has relatively large flowers and blooms in summer.

ZEUXINE (*Zeuxine strateumatica*) This is one of the exotic orchids which became naturalized in this country. It was introduced accidentally several years ago and has now spread to various parts of Fla., showing weedy characteristics. A native of Asia, it was probably introduced with seeds of centipede-grass. Flowers are small, white, or yellowish with a yellow lip.

[25]

multi colored

white orange lip

Corallorhiza Wisteriana Zeuxine strateumatica

white   white

Habenaria nivea - H. Blephariglottis

HABENARIAS (*Habenaria*) This is a genus of terrestrial, green herbs with an erect and usually leafy stem. Flowers are small to medium-sized with a slender, cylindrical spur, borne in inconspicuous or showy racemes. The most beautiful Habenarias have lips more or less deeply fringed into hair-like divisions. Some of these are known as fringed-orchids. The most frequent and widespread of these is the yellow-fringed (*H. ciliaris*) which occurs from the coast to the mountains. Very similar to this is the white-fringed (*H. Blephariglottis*) except for its pure white color. This is widely distributed but is more frequent in the Coastal Plain than further inland. Both bloom during the summer.

SNOWY-ORCHID, WHITE REIN-ORCHID (*Habenaria nivea*) This is smaller than the white-fringed, with the lip uppermost, not fringed and with a long curved spur. It blooms earlier than the above and occurs in similar habitats in the Coastal Plain. Fla. to Tex. Ark. and N.J. It is sometimes called "bog-torches." *H. strictissima* is a tropical orchid, occurring throughout the W.I., C.A. and Mex., extending into s. Fla. in the var. *odontopetala* which means toothed-petal. It occurs in rich, wet soil of various situations.

greenish   Lavender

Habenaria clavellata   H. psyco des

LITTLE CLUB-SPUR ORCHID (*Habenaria clavellata*) This is one of the commonest orchids in the e. U.S. but is often overlooked having inconspicuous flowers. It has greenish or yellowish-white flowers with a conspicuous club-shaped spur.

SMALL PURPLE-FRINGED ORCHID (*Habenaria psycodes*) The purple-fringed orchids are more frequent in upland districts, some occurring only at higher elevations. The small purple-fringed occurs at medium altitudes, frequenting meadows, sedge-swamps, and other wet situations. It has a northern distribution but ranges as far south as Ga. Another interesting habenaria of the South is the southern rein-orchid (*H. flava*) with yellow flowers partly hidden behind the long, pointed bracts. It is typically a Coastal Plain plant but may extend into other provinces. Infrequent but often locally abundant. The most aquatic of our orchids is the water-spider orchid (*H. repens*) which grows in wet places and sometimes occurs on floating mats of decaying vegetation. It has small, greenish flowers with a strongly reflexed, 3-parted lip.

after
B. Ames.   X   greenish

Habenaria repens   H. strictissima var. odontopetala

LIZARD'S TAIL FAMILY (Saururaceae)

LIZARD'S-TAIL (*Saururus cernuus*) From early spring in the southern parts of our area and later northward, the graceful, drooping, tapering, white racemic-spikes of the lizard's-tail decorate marshes, swamps, roadside ditches, and borders of streams throughout the eastern states. Because of its extensively creeping rootstocks, it forms thick colonies among the sedges, rushes, and cattails growing in similar habitats. The small flowers and the rootstocks are very aromatic. Fla. to e. Can. and w. to Mich. Suitable for garden pools and fountains.

Saururus cernuus - white

BAYBERRY FAMILY (Myricaceae)

CANDLE-BERRY, WAX-BERRY, WAX-MYRTLE (*Myrica cerifera*) The wax-myrtles are shrubs which sometimes grow to tree size. Their flowers are not especially attractive as they are borne in scaly catkins, or aments, as in willows, oaks, and certain other woody plants. The sexes are separate so only plants bearing female flowers bear fruit. The small berries exude a wax which turns them white at maturity. The leaves as well as the berries are pleasantly fragrant. Frequent in swamps and hammocks throughout the Coastal Plain but most common near the coast. Occasionally in higher ground and extending into the Piedmont. Fla. to Tex., Ark. and n. to s. N.J. Four other species of myrtle occur in the South.

greenish catkins

Myrica cerifera - fruit gray

BUCKWHEAT FAMILY (Polygonaceae)

DOG-TONGUE, WILD-BUCKWHEAT (*Eriogonum tomentosum*) This dog-tongue is distinguished by its whorled leaves which are green above and white-wooly beneath. The small, hairy flowers are white to pinkish and are borne in clusters with an involucre of leaf-like bracts at the base. Frequent in dry pinelands and sandhills, Coastal Plain, Fla. to Ala. and S.C. *E. floridanum*, restricted to c. Fla., has alternate leaves and small involucral bracts. *E. Alleni*, occuring locally on shaded slopes and barrens in Va. and W.Va., has a leafy inflorescence and nearly smooth, yellow sepals. This is especially suitable for wild flower gardens and arrangements.

[27]

yellow

pink and

Eriogonum Alleni   E. tomentosum - white

JOINT-WOOD, OCTOBER-FLOWER (*Polygonella polygama*) The copiously branched plants of *Polygonella* covered with white, persistent flowers are attractive on open shrubby pinelands and sandhills in the autumn when most plants are through blooming. This species is an annual up to 2′ tall. It is a very decorative plant for dried bouquets and arrangements. Ranges in the Coastal Plain from Fla. to Miss. and N.C. Four other species of *Polygonella* occur in the South.

white tinged pink

Polygonella polygama.- P. gracilis

SMARTWEEDS, GANDER-GRASSES (*Polygonum*) These are familiar plants to anyone who has wandered about in marshy places in late summer and autumn. The rather small flowers are borne in spikes which in the more showy ones are dense and dark-pink or reddish. Because of their rank and weedy nature, the attractiveness of the native smartweeds is not generally appreciated. One species (*P. orientale*), a tall, hairy annual, known as "princess-feather" or "kiss-me-over-the-garden-gate," has been introduced from Eurasia and is spreading to some extent from cultivation. Two of our most ornamental native species are *P. coccineum,* a perennial with rose-colored flowers, and *P. pensylvanicum,* an annual called "pink-weed," with pink or light-purplish flowers.

pinkish *rose*

Polygonum pensylvanicum-P. coccineum

SEA-GRAPE, GRAPE-TREE, PLATTER-LEAF (*Coccoloba uvifera*) A shrub or small tree with crooked stems and orbicular, cordate leaves forming thickets on dunes and coastal hammocks in s. Fla. and throughout the W.I., the sea-grape adds a distinctive feature to the picturesqueness of tropical seaside vegetation. The reddish, berry-like fruits, ½″-¾″ in diameter, are edible and are used for jelly when enough remain after certain crabs come up from the ocean at night and devour them. The small flowers are borne in slender spikes above the leaves. It should be planted more than it is as an ornamental and the leaves are uniquely attractive in arrangements.

[28]

x

Coccoloba uvifera - whitish-red fruit

AMARANTH FAMILY (Amaranthaceae)

COTTON-WEED (*Froelichia floridana*) This interesting and distinctive plant is a tall annual or biennial which gets its common name from the dense hairiness of the under side of the leaves and the outside of the flowers. The many small, greenish-white flowers are borne in short, dense spikes or heads. The long, freely branching tomentulose stems sprawl over the ground. Pinelands, sandhills, and old fields, Coastal Plain, Fla. to Miss. and sporadically to N.C. Globe-amaranth or bachelor's-button (*Gomphrena dispersa*) is an exotic perennial herb from the tropics which used to be a favorite flower in old-fashioned gardens. Since then, it has escaped and become naturalized in the Gulf states. Small bracts between inconspicuous flowers give a variety of colors to the heads. Roadsides and waste places, Fla. to Tex.

Froelichia floridana Gomphrena dispersa

GOOSEFOOT FAMILY (Chenopodiaceae)

GLASS-WORTS, SAMPHIRES (*Salicornia*) Along the margins of ocean inlets, these curious creeping or erect, jointed, leafless, fleshy plants grow, called glass-worts from the crunching sound they make when stepped on. The perennial ones (*S. virginica*) have creeping stems with erect branches, while the annual kinds (*S. Bigelovii* and *S. europea*) grow erect like miniature trees. Tiny fleshy flowers are imbedded in the joints. These and the following 2 are edible and have been found to be of survival value when cooked.

CARPET-WEED FAMILY (Aizoaceae)

SEA-PURSLANE (*Sesuvium Portulacastrum*) This fleshy annual with smooth, opposite wedge-shaped leaves is a common plant of sandy beaches of harbors and estuaries. Small apetalous flowers are borne in the upper axils, with many stamens and sepals bordered with white. The capsules open by a conical lid exposing numerous shiny, black seeds. *S. maritimum* differs from the above mainly in having only 5 stamens.

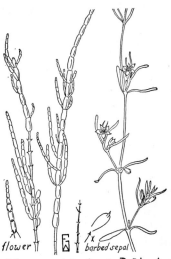

flower *barbed sepal*

Salicornia virginica Sesuvium Portulacastrum

SALTWORT FAMILY (Batidaceae)

SALTWORT, BEACHWORT (*Batis maritima*) A low, fleshy shrub with spreading and creeping stems rooting at tips forming patches of pure stands. The pale-green foliage is very striking. Unisexual flowers are borne in short, axillary, cone-like spikes. Beach sand, Fla. to Tex. and N.C.

[29]

fruit *staminate flower* *pistilate*

Batis maritima *edible*

white or pink

Rivina humilis   red berries

POKEWEED FAMILY (Phytolaccaceae)

BABY-PEPPER, ROUGE-PLANT (*Rivina humilis*) A relatively small, often partially woody plant with stems 4″-30″ tall, or sometimes with vine-like stems. Small, 4-petaled, white to pink flowers are borne in racemes. Leaf blades are ovate, lanceolate, or elliptic and undulate, smooth or hairy. The most attractive part of this plant is the small, red berries. Seen all year throughout its range from Fla. to Tex. and Ark. Sometimes grown as an ornamental in northern greenhouses. Related to our common pokeweed or pigeon-berry (*Phytolacca americana*) widely distributed in e. U.S. as well as *P. rigida* of the Gulf states.

Stellaria pubera   white

PINK FAMILY (Caryophyllaceae)

GREAT-, GIANT- or STAR-CHICKWEED (*Stellaria pubera*) When you tramp the woods on river bluffs in the early spring and find a low, bunchy plant with white star-shaped flowers which seem to have 10 petals, it is undoubtedly the native giant-chickweed. There are really only 5 petals but they are so deeply cleft that there appear to be 10. The flowers are further ornamented by about 10 stamens with red anthers. This is related to the common chickweed (*S. media*) a pestiferous winter annual in our gardens. The giant-chickweed grows in rich rocky woods of the Piedmont and mountains, and ranges from n. Fla. to Ala. and n. into Can.

red   pink

Silene virginica   *S. carolinana*

INDIAN- or FIRE-PINK (*Silene virginica*) This is the showiest of our native pinks, having a relatively large flower with deep red or crimson petals. It is a slender, hairy, weak-stemmed perennial growing in rocky ground especially on river bluffs from the Piedmont to the mountains. The petals are unique in being notched at tip and having a scale-like appendage at the base. Ranges from Ga., northward. Another attractive pink, which is more typically southern, is *S. Baldwinii* (color page facing 32) with white to pink petals. It is restricted to the Coastal Plain and Piedmont of Fla. and Ga.

[30]

BUTTERCUP FAMILY (Ranunculaceae)

BLACK SNAKE-ROOT, BLACK-COHOSH, BUGBANE (*Cimicifuga racemosa*) (color page facing 32) The traditional virtues of this stately herb for curing snake bites and driving away "bugs," as those of many another herb, has faded in the light of modern medicine; but the admiration of its long, white, racemic, candle-like spikes on our wooded hillsides during late spring and early summer remains and grows. The stem is from 3'-7' tall and the large basal leaves are ternately compound. The individual flowers with their numerous stamens with slender filaments look like small brushes. Lower Piedmont to mountains from Ga. to Tenn. and N.C. and has spread from cultivation northward.

Cimicifuga racemosa - white

AMERICAN COLUMBINE (*Aquilegia canadensis*) (color page facing 48) The name Columbine is supposed to have come from a fancied resemblance of the flowers to doves, and *Aquilegia* from the talons of eagles suggested by the nectar-spurs of the flowers. Such imagination reminds one of conceptions of the constellations of the stars! The interesting form and beauty of color of this flower is unsurpassed and is a striking example of the adaptation of flowers to insect pollination. It keeps well when cut and is successfully cultivated from seed throughout its range and beyond. It varies considerably in the length of sepals, length and shape of spurs, form of leaves, and general habit. Woods and rocky banks of Piedmont and mountains, from Ga. northward.

Aquilegia canadensis-scarlet & yellow

FALSE-BUGBANE, TASSEL-RUE (*Trautvetteria caroliniensis*) (color page facing 33) Although extending along river bluffs as far down as the Coastal Plain, it is on the borders of mountain streams and in the mist of waterfalls that the false-bugbane is at home. The pure white sepals and numerous club-shaped stamens make up the showiness of the flowers which have no petals. The relatively large, palmately-lobed leaves and the white brush-like flowers are the distinguishing marks of this distinctive and attractive plant. W. Fla. to Mo. and Pa.

[31]

Trautvetteria caroliniensis - white

Amsonia Tabernaemontana

Baptisia australis

Baptisia alba

Chamaelirium luteum

Malus coronaria

Erigeron pulchellus

Galax aphylla

Cypripedium Calceolus

R (Azalea) sp.

Biltia Vaseyi

Trillium cernuum

v. parviflorum

C. pubescens

Rhod(Azalea) calendulaceum

Viola cuccullata

Uvularia sessilifolia

Gillenia trifoliata

Cypripedium
acaule

R. carolinianum
var. alba

Viola canadensis

Menziesia
pilosa

R.(azalea) nudiflorum

Clintonia umbellulata

Galax leaf

Iris versicolor

R. carolinianum

Phlox ovata

Rhododendron catawbiense
rose form

Orontium aquaticum

Robinia hispida

Iris cristata

Tiarella cordifolia

R. catawbiense

Viola Langloisii

Silene Baldwinii

Lonicera flava

Houstonia
serpyllifolia

SPRING WILD FLOWERS OF THE CAROLINA MOUNTAINS

SPRING WILD FLOWERS OF THE CAROLINA MOUNTAINS

The lady slipper orchids are too rare to be picked (except where found in great abundance) nor are they happy in the wild-flower garden. All others last well and can be picked sparingly; they are excellent for the rockery or wild gardens.

Lilies and Other Summer Wild Flowers
of the Southern Mountains

Aruncus dioicus   Amianthium Muscaetoxicum
Lilium Grayi
Trautvetteria caroliniensis
Tradescantia virginiana
Diervilla sessilifolia
Lilium canadense
Xerophyllum asphodeloides Baptisia tinctoria
Lilium superbum Thermopsis caroliniana
L. Grayi
Viburnum acerifolium Allium tricoccum
Myosotis scorpioides

LILIES AND OTHER SUMMER WILD FLOWERS
OF THE SOUTHERN MOUNTAINS

These glorious lilies and other flowers are the most outstanding summer flowers of the southern mountains and the upper Piedmont. It is breath-taking to see the Turk's-cap (*Lilium superbum*) in moist, shaded dells as tall as one's head, with 30-50 flowers in one inflorescence. The Carolina lily (*L. Michauxii*) is very similar but has lighter leaves which are broadest beyond the middle. In Canada lily (*L. canadense*), and its close relative the Gray's lily (*L. Grayi*), the perianth is not as much recurved as in the above two. The wood-lily (*L. philadelphicum*), with erect flowers, is rarely seen in the southern mountains. Most conspicuous among the white flowers is goat's-beard (*Aruncus dioicus*) with its sprays of unisexual flowers, the fly-poison flower (*Amianthium muscae-toxicum*) with its racemic-spike similar to that of the mountain asphodel (*Xerophyllum asphodeloides*), and false bugbane (*Trautvetteria caroliniensis*) of margins of mountain brooks and water falls. Among the few yellow flowers, the most conspicuous are Aaron's-rod (*Thermopsis caroliniana*), which is often cultivated, and yellow-indigo (*Baptisis tinctoria*). *Allium tricoccum* is the wild-leek, or "ramp" eaten by the mountain people. Sandy brooks are the home of forget-me-nots (*Myosotis scorpioides*). All of these and many more may be seen at the Pearson's Falls Wild Flower Preserve, near Tryon, N. C., maintained by the local women's garden club.

Anemone caroliniana - purple or white

CAROLINA ANEMONE (*Anemone caroliniana*) This southern representative of the genus *Anemone* was discovered and named by the early southern botanist, Thomas Walter (1740-1789) of Charleston, S. C., who published one of the earliest accounts of southern plants. It is a tuberous-rooted perennial with a solitary flower on a single flowering stem. The flowers are white to roseate, rarely deep violet, with 6-20 sepals. It grows in dry, rich, or stony soil in various provinces and should be successfully cultivated from seed in rock gardens. *Anemone virginiana*, commonly called "thimble-weed," is a coarse, hairy plant with rather large, compound stem leaves, which ranges from Ga. northward in e. U.S.

greenish white — white-or pinkish

Anemone virginiana Anemonella thalictroides

RUE-ANEMONE, WIND-FLOWER (*Anemonella thalictroides*) This is another of our early spring flowers, associated with hepatica and trout-lily. It is not a true *Anemone* and the leaves resemble those of *Thalictrum* or meadow-rue, hence the first common name. It is a low perennial with one compound basal leaf to each plant and a whorl of leaf-like bracts below the group of white flowers on long, slender pedicels. It is easily transplanted to wild flower gardens. Moist woods, thickets, and stream banks, in various districts but confined to uplands, from n.w. Fla. northward.

deep blue — blue

Delphinium carolinianum · D. exaltatum

DELPHINIUM or CAROLINA LARKSPUR (*Delphinium carolinianum*) Although the flowers of the common garden larkspur resemble more the head of a rabbit than a dolphin, the latter won out in the selection of a name for these plants. The wild Delphiniums resemble closely those which are cultivated, but the flowers are smaller and more delicate both in outline and color. The Carolina delphinium has usually a simple stem 1'-2' tall with several lower leaves deeply 3- to 5-parted. It grows in rocky woods, sand-hills, and barrens in various provinces from Fla. to Tex. and n. to Mo., Ky., and Va. The tall delphinium (*D. exaltatum*) differs from the above in larger size of all its parts. S. Ala. to Ohio and Pa.

[34]

Monk's-Hood, Wolfsbane (*Aconitum uncinatum*) The wolfsbanes or monk's-hoods have for long been known to possess a powerful poisonous narcotic in their roots and received the name wolfsbane from their use in poisoning wolves. The other name alludes to the resemblance of the flowers to a hood. This beautiful species is an ascending or reclining perennial with thick, divided leaf blades. It is not frequent nor abundant where found. Piedmont and mountains, Ga. to Ala. and locally northward.

Aconitum uncinatum   blue

Hepatica or **Liver-Leaf** (*Hepatica americana*) This is an early spring flower which blooms with the trout-lily, spring-beauty, and Anemones. The hairy, bluish-tinted flower-buds may be seen pushing up between the fallen leaves of the previous autumn before heavy frosts have vanished from the land. The brownish, 3-lobed leaves have a fancied resemblance to the liver, and, therefore, according to the "doctrine of signatures" of the Middle Ages, must be good for liver ailments. Various provinces in rocky woods and on stream banks, n. Fla. to Can. *H. acutiloba,* the only other hepatica of e. N.A., has more pointed leaf lobes, lighter-colored flowers and is restricted to higher altitudes.

Hepatica americana-bluish to white

Leather-Flowers (*Clematis*) Our native species of *Clematis* with relatively large, bell-shaped flowers and thick, leathery sepals are generally called "leather-flowers" but may also have other local names. Most of them are woody vines but a few are low herbs called dwarf clematis. One of these is *C. Baldwinii*, called pine-hyacinth, which is restricted to peninsular Fla. The flowers are bell-shaped with strongly recurved, purplish, crisped sepals. Of the woody vines, one of the most attractive is curly- or marsh-clematis (*C. crispa*). The calyx tips are pale-blue and crisped on the margins. Range: Fla. to Tex. and Va. A similar but less attractive plant is *C. reticulata* which ranges in the Coastal Plain from Fla. to Tex. and n. to Va.

[35]

Clematis Baldwinii C. reticulata-mauve

1

1

cream
with purple 2 ⊡ 3.

1Asimina obovata 2 A.reticulata 3. fruit

CUSTARD-APPLE FAMILY (Annonaceae)

This is mainly a tropical family which furnishes several valuable fruits, such as cherimoya, soursop, sugar-apple, and others upon which the inhabitants of tropical islands in both the Atlantic and Pacific largely depend for their subsistence. The only species which range into temperate regions are *Asimina triloba,* a small tree, known as "paw-paw," which is edible, and *A. parviflora,* or "dwarf paw-paw" called "possum-simmon" in certain sections of the South.

PAW-PAWS, FLAG PAW-PAWS, DOG-APPLES (*Asimina*) *Asimina reticulata* is a shrub up to 3', the shoots tawny and tomentulose; flowers white, arising from axils of deciduous leaves of the preceding season, blooming in winter and spring. Fruits elliptic, 1"-2½" long. *Asimina obovata,* which is locally called "flag paw-paw," has leaves leathery, broadened upward, appearing before the flowers which are white to cream. Winter and spring. Fla. *Asimina angustifolia* grows up to 4' tall with long, narrow leaves, appearing before the flowers which are white to cream, 2"-3" long. The petals grow after the flower opens. Pinelands, Fla. to Ga. *Asimina pygmaea,* a shrub, 1'-2' tall, erect or prostrate, has oblong leaves, appearing before the flowers which are at first greenish-white, changing to reddish-purple, 1"-2" long. Spring and summer. Pinelands, Fla.

cream
with purple

Asimina angustifolia   A. pygmaea

MOONSEED FAMILY (Menispermaceae)

CORAL-BEADS, RED-MOONSEED, CAROLINA MOONSEED (*Cocculus carolinus*) The moonseed family is so named from the crescent-shaped seeds of some of its members. All are herbaceous, perennial, mostly tropical vines. Coral-beads is the most attractive of our native vines and has small greenish flowers and coral-red berry-like fruits (drupes) about the size of a small pea. It is an attractive vine and should be cultivated more than it is. Fla. to Tex., n. to s.e. Va., and Kans. The common moonseed (*Menispermum canadense*) of northern distribution occurs as far south as Ga. It differs from coral-beads in leaf characters and has bluish-black drupes.

[36]

blue drupe
red drupe   white flowers
Cocculus carolinus   Menispermum canadense

MAGNOLIA FAMILY (Magnoliaceae)

MAGNOLIAS (*Magnolia*) Of the several species of *Magnolia* occurring in the South, the evergreen loblolly- or southern-magnolia (*M. grandiflora*) (frontispiece) is the best known being planted far beyond its natural range of Fla. to La., s. Ark. and s.e. N.C. The large handsome, creamy white flowers are lemon-scented. Sweet- or swamp-bay magnolia (*M. virginiana*) characterizes the so-called "bay-lands" of the s. Coastal Plain. Both the white flowers and the foliage are pleasantly fragrant. A southern variety (*australis*) has silky-white underside of leaves and branchlets. Great-leaved magnolia (*M. macrophylla*) with its northern range limits in N.C. is cultivated successfully much farther north.

Magnolia virginiana (glauca)   white

TULIP-TREE, YELLOW-POPLAR (*Liriodendron Tulipifera*) This is one of our most beautiful forest trees, having a straight trunk with a tight, attractive bark. It may attain a height of 130' with a diameter of up to 10'. The botanical name *Liriodendron,* meaning "lily" or "tulip-tree," alludes to the large tulip-like flowers with greenish-yellow petals marked with orange, as does also its specific name *Tulipifera*. It commonly blooms in late April and early May in the South. Although it ranges as far north as s. Ont. it seems to be typically a southern tree. It furnishes the valuable timber called "poplar" or "white-wood." It is the only species occurring in N.A. but has a close relative in China.

Liriodendron Tulipifera-yellow-gr. & orange

PURPLE-ANISE, STINK-BUSH, ANISE-TREE (*Illicium floridanum*) The Illiciums are among the most attractive evergreen shrubs of the subtropical South. Related to the Magnolias, they are aromatic and the flowers have 3-6 sepals and petals in 3 to several series, yellow or red and narrow. Corollas—as well as the fruits—appear star-shaped. Purple-anise has dark-red petals, but unfortunately they smell like decaying fish. Swamps and low hammocks, Coastal Plain, Fla. to La. Star-anise (*Illicium parviflorum*) has small yellow flowers and acuminate leaves. Low woods and swamps, Coastal Plain, Fla. to Ga. Both are suitable for borders and foundations and for parks and roadsides. Hardy as far north as n. Miss.

maroon

Illicium floridanum "Purple-anise"

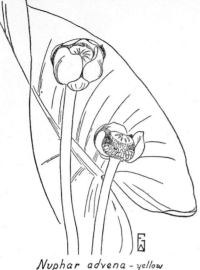

Nuphar advena - yellow

yellow

white

Nymphaea flava-yellow N. odorata-white

Nelumbo lutea - pale yellow

WATER-LILY FAMILY (Nymphaeaceae)

YELLOW POND-LILIES (*Nuphar*) The yellow pond-lilies are easily distinguished from other water lilies by their yellow, depressed-roundish flowers and thick, bluntish petals. They bloom in late spring and continue blooming during the summer months. One of the most wide-spread is spatter-dock, bonnet or cow-lily (*N. advena*) with usually erect leaves and large, heart-shaped blades. Margins of tidal waters, ponds, and streams, Fla. to Tex. and northward. *Nuphar sagittifolia*, called "arrow-lotus," which grows mainly in slow-moving streams of the Coastal Plain, has long, broad strap-shaped floating leaves. It is restricted to e. S.C. and N.C.

WATER-LILIES, WATER-NYMPHS, POND-LILIES (*Nymphaea*) One of the delightful sights as one travels in the lower Coastal Plain is the frequent cypress and tupelo-gum ponds covered with patches of white, pinkish, blue to violet, and yellow water-lilies. One of the most widely distributed of these is the white to pinkish "alligator-bonnet" or "star-lotus" (*N. odorata*) which ranges from Fla. to La. and n. to e. Can. Of more local distribution is the blue water-lily (*N. elegans*), restricted to ponds and ditches of the Big Cypress Swamp of Fla. and in s. Tex. The distinctive yellow water-lily or sun-lotus (*N. flava*) is restricted to peninsular Fla. All are used in ponds and fountains.

LOTUS FAMILY (Nelumbonaceae)

AMERICAN LOTUS, WATER-CHINQUAPIN, DUCK-ACORN (*Nelumbo lutea*) The American lotus has not yet entered into our legendary romance, art, religion, and economic use as has its close relative, the sacred lotus of India. The latter should not be confused with the "lotus" referred to in the Greek legend of the lotus-eaters who surrendered to a dreamy indolence. Our lotus bears large, pale-yellow flowers of many petals. The nuts or "seeds" are imbedded in a funnel-shaped receptacle which is frequently gilded and sold in florist shops. The large, circular leaf blades are borne on stout petioles which raise them above the water. Ponds, lakes, and slow streams of the Coastal Plain and Mississippi Valley. Fla. to Tex. and northward to e. Can. Often planted in ponds.

[38]

POPPY FAMILY (Papaveraceae)

BLOOD-ROOT, RED-PUCCOON (*Sanguinaria canadensis*) One of the characteristics of the poppy family is the production of colored juices or "saps" so that when bruised they "bleed." This juice is the source of opium, obtained from the young capsules of the opium-poppy. In blood-root, the juice is red and is especially abundant in its short rootstocks. The white flower with its two evanescent sepals and several petals is borne on a naked stalk beside a waxy, lobed leaf which precedes it in spring. The petals soon drop after opening of the flowers. Rich woods and narrow, rocky stream bluffs, Fla. to Ark. and n. in e. U.S. to Can.

Sanguinaria canadensis - white

PRICKLY-POPPIES and THORN-APPLES (*Argemone*) Waxy, leafy-stemmed, usually spiny annual or biennial herbs with flowers which resemble those of poppies, but with sepals and capsules more or less spiny. Originally of the arid Southwest, they are rapidly spreading to eastern areas. The Mexican or yellow-thistle occurs on roadsides and in waste places from Fla. to Tex. and n. to Tenn., Penn., and Mass. Summer and fall and all year further south. The decorative white- or Caro-poppy is found in similar situations but is not as widely distributed. Fla. to Tex., Mo. and N.C. Easily grown from seed.

white   *yellow*
Argemone alba   *A. mexicana*

BARBERRY FAMILY (Berberidaceae)

UMBRELLA-LEAF (*Diphylleia cymosa*) The umbrella-leaf has in common with may-apple (*Podophyllum peltatum*) and twin-leaf (*Jeffersonia*), also of this family, two large leaves between which the flowers are borne. However, it differs from these two in having a cyme of several flowers while they have only one. The leaves are large, 2-cleft and lobed with pointed lobes. The flowers have 6 white petals. Fruits consist of attractive blue berries about ½' in diameter. It is a cool-climate plant and grows along brooks of the Blue Ridge from Ga. to Va. An excellent decorative and interesting plant for cool, shaded situations.

[39]

berry   *stamen X*
Diphylleia cymosa - white

white   x½

Dicentra Cucullaria

FUMITORY FAMILY (Fumariaceae)

DUTCHMAN'S-BREECHES (*Dicentra Cucullaria*) For this unique, delicate flower the common name is happily chosen for the 2-spurred waxy-white flowers on a slender, arching stalk provoke a smile of recognition in their resemblance to tiny, baggy breeches hung on the line to dry. An early-blooming perennial having a cluster of grain-like tubers crowded together in a scaly bulb from which grow the finely-cut, delicate, waxy leaves, and the slender, naked flowering stem bearing the waxy-white flowers with cream-colored crests. It grows in rich woodsoil in the Piedmont and mountains from Ga. northward and in parts of n.e. U.S. and e. Can. Although sometimes abundant locally, it is scattered in distribution. Easily grown in rich, shaded situations in wild flower gardens.

white *Dicentra canadensis*

SQUIRREL- or TURKEY-CORN (*Dicentra canadensis*) Leaves of squirrel-corn are very similar to Dutchman's-breeches, but the flowers are heart-shaped at the base instead of spurred as in the latter. The color of the flowers is a greenish-white, tinged with rose and the fragrance is of hyacinths. The underground shoots bear fleshy, yellow, grain-like, thickened leaf-bases which give it its common names. In rich woodsoil, from N.C., Tenn., and Mo. n. to e. Can.

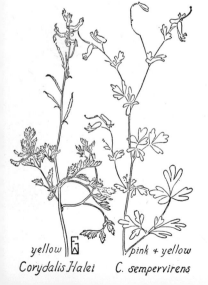

yellow   pink + yellow

Corydalis Halei   *C. sempervirens*

WILD-FUMEROOTS, CORYDALIS, SCRAMBLED-EGGS (*Corydalis*) The plants included under the genus *Corydalis* differ from the above 2 in having only one spur on the flowers which are borne on leafy stems. The color of the flowers may be pink, purple, or shades of yellow. The foliage, quite similar in all of them, is dissected and more or less a waxy, light-green. Pink- or pale-corydalis (*C. sempervirens*) has an erect stem and grows on rock exposures and cliffs at higher altitudes. The others are mostly weak-stemmed, diffuse plants with yellow flowers growing in rich soil. One of the latter found only in the lower South is *C. Halei* which grows in sandy soil in the Coastal Plain from Fla. to Tex. Three others extend farther north.

[40]

MUSTARD FAMILY (Cruciferae)

GOLDEN LEAVENWORTHIA (*Leavenworthia aurea*) The Leavenworthias are small, N.A. annuals, glabrous and often stemless with rosettes of lyre-lobed leaves and short, 1- to few-flowered scape-like peduncles. Named for Dr. Melines Conklin Leavenworth (1796-1862), a southern botanist. The golden leavenworthia is distinguished from 3 other species occurring in the southern states by its relatively large and blunt-lobed leaf segments, its white or purplish flowers with a yellow base, and short pods with a short, slender beak. It inhabits rocky, often calcareous soil in various situations except in the Coastal Plain. It is southern in distribution, ranging from Ala. to Tex., Ark. and Tenn.

white or purplish-yellow base
Leavenworthia aurea

SEA-ROCKET (*Cakile Harperi*) The sea-rockets are fleshy annuals growing on ocean beaches close to the upper limits of the waves where few other plants can grow. White to pale-pink flowers are borne in racemes and the fruits are spindle-shaped, fleshy pods with a cross partition and one seed in each half. At maturity, the upper half separates from the lower which later falls from the pedicel. The seeds are not shed from these half-fruits which undoubtedly protect them from sea water which serves as agent in their dispersal. The mustard flavored leaves, eaten raw or cooked, helped G.I.'s stranded on barren beaches to survive.

edible
Cakile Harperi - white, pink or purple

WAREA (*Warea amplexifolia*) Wareas are slender herbs with tooth-edged leaves and white, rose, or purple flowers. Only 4 species are known and 3 of these are restricted to the sandhills of Coastal Plain Fla. *W. amplexifolia* is from 1'-3' tall with white flowers which fade purple. The loose racemes lengthen during the blooming period, while the pods of the lower flowers lengthen in persistent perianths. Keeps well when cut.

CAPER FAMILY (Capparidaceae)

CATCH-FLY (*Aldenella tenuifolia*) Related to the cultivated spider-flower (*Cleome spinosa*), catch-fly has a sticky substance on its stems which catches ants and other small insects. The white flowers have 2 large, stalked petals and 2 small ones which are strongly reflexed exposing the stamens and a curved pistil. Sandy soil, Fla. to Miss.

lavender   *white*
Warea amplexifolia-Aldenella tenuifolia

Plants which imprison insects and other small animals exert a peculiar effect upon the human imagination. This fact has given rise to such fantastic stories as "The Man-eating Tree of Madagascar" and others. According to Francis Ernest Lloyd, author of *The Carnivorous Plants* (1942), at least 450 species of plants, representing 15 genera and 6 families, are carnivorous. These are in general not restricted to any particular region but are widely distributed over the earth, although in limited areas they may be associated with certain soil types correlated with certain climates. The same divergence is found in genetic relationships; for, although certain plant families have a larger share of carnivorous forms than others, the carnivorous habit seems to have developed in several distinctly unrelated groups. To be an insectivorous plant, there must be some evidence of utilization of the animal by the plant, but there are many examples of plants which catch insects, such as catchflies (*Silene*), in which there is no indication that the insect is being used. What the "purpose" of this latter situation may be, we do not know; but we are almost as much in the dark regarding the function, or its importance, where the animals are digested and absorbed. The methods of catching insects and other small animals have been classified by Lloyd into the passive and active. To the first class belong the pitfalls furnished by pitcher-plants and the "lobster pots" of *Genlisea* and the flypaper type of some which in the sundews is combined with an active phase. The active types are represented by the remarkable adaptations of trap-like mechanisms found in the Venus'-flytrap (*Dionaea*) of N.C. and S.C. and *Aldrovanda*, an aquatic plant of s. Europe. Perhaps the most remarkable of all are the cage-traps of the aquatic bladderworts (*Utricularia*). In addition, there are lures of various types, such as excretions in sundews, bright colors as in the glands of the Venus'-flytrap, nectar and odors in others, and the translucent "windows" of the hooded pitcher-plant. The s. Coastal Plain has a remarkable assortment of insectivorous plants, the most common of which are pitcher-plants, sundews, bladderworts, and butterworts (*Pinguicula*). More restricted in distribution is the most interesting of them all, the Venus'-flytrap. It is not only highly specialized in its trap-like leaf, but has a high degree of sensitivity and active motion. When acting at its best, it may even give one the impression of aggressiveness.

VENUS'-FLYTRAP FAMILY (Dionaeaceae)

VENUS'-FLYTRAP (*Dionaea muscipula*) This remarkable plant consists of a prostrate stem and a rosette of leaves with winged petioles, a trap-like blade with marginal bristly hairs and a prominent midrib on the under side. The upper side is covered with glands, ranging in color from light-red to scarlet. In addition, it has usually 3 "trigger hairs" on each upper half of the blade which, when bent by moving objects such as insects or spiders, set a mechanism in motion which quickly closes the blade, imprisoning the moving object. Even small frogs may be caught. After the initial closure, the trap closes more tightly and the glands secret a clear, sticky liquid which digests the softer parts of the victims. Restricted to s.e. N.C. and n.e. S.C.

Dionaea muscipula white

PITCHER-PLANT FAMILY (Sarraceniaceae)

PITCHER-PLANTS (*Sarracenia*) The pitcher-plants may be classified into those with relatively short, decumbent leaves, and those with longer upright ones. To the former group belongs the purple pitcher-plant (*S. purpurea*) of wide distribution in e. N.A. Another one of this group with more limited distribution is the parrot pitcher-plant (*S. psittacina*) in which the hood over the opening of the pitcher resembles a parrot's head. This odd-looking plant has purplish flowers and is restricted to n. Fla., s. Ga. to La. Of the upright type, the hooded pitcher-plant is unique in having numerous transluscent spots on the back of the hood and the upper portion of the body. S.e. N.C. to Okeechobee region of Fla. Hybridization in pitcher-plants is not uncommon.

yellow   purplish
Sarracenia minor   *S. psittacina*

SUNDEW FAMILY (Droseraceae)

SUNDEWS (*Drosera*) The sundews are low, insectivorous plants with a rosette of leaves with blades varying from circular to spathulate or thread-shaped, covered on the upper side with numerous gland-tipped hairs which secrete a sticky substance which glistens innocently in the sunshine. However, when small insects or spiders crawl over the leaf, they not only bog down in the sticky fluid but the gland hairs curl over them, like fingers gripping a butterfly, and, soon after exhaustion and death, they are digested and absorbed. Small, white or pink flowers are borne at the end of a slender stalk. They grow in bogs or boggy soil most frequently in the Coastal Plain. Five species occur in the South.

white

Drosera brevifolia filiformis·intermedia

insect caught on sticky tentacles

white
coral anthers · lavender
Sedum ternatum · S. pusillum

SEDUM or ORPINE FAMILY (Crassulaceae)

MOUNTAIN- or THREE-LEAVED STONECROP (*Sedum ternatum*) The commonest of the Sedums in the Piedmont and the mountains is this low, spreading plant rarely over 8″ tall, characterized by a mosaic rosette of wedge-shaped, flat, opposite leaves in whorls of 3. Numerous 4-petaled, white to lavender flowers with orange to red anthers are borne on a 3-forked, flat-topped inflorescence in early spring. On rocky stream banks in various provinces except in the Coastal Plain. Ga. n. to the northern states. Excellent for rock gardens and ground cover under trees. Puck's-orpine (*S. pusillum*) is a small plant, 2″-5″ tall with wedge-shaped leaves and numerous white to pink flowers and short-beaked fruits. Piedmont and mountains, Ga. to N.C.

SAXIFRAGE FAMILY (Saxifragaceae)

ALUM-ROOTS, ROCK-GERANIUMS (*Heuchera*) The alum-roots are typically inhabitants of rocky ground, blooming in spring or early summer. They have basal rosettes of geranium-like leaves and tall, slender flowering stalks with panicles of relatively small flowers. In the common alum-root (*H. americana*) the flowers are not especially attractive but the rosettes of blunt, unevenly-toothed leaves with cordate bases have an ornamental value in rock gardens. Common and easily transplanted. Extremely variable with many varieties. Uplands, n. Ga. to e. Can. Among other species, is the small-flowered alum-root (*H. parviflora*) which differs from the above in having longer and usually white petals with long claws. Base of cliffs and shaded rocky situations at higher elevations. Ga. to Va. and Ill.

Heuchera americana - H. parviflora · white

FOAM-FLOWER, FALSE-MITREWORT (*Tiarella cordifolia*) This is the showiest and most attractive of our spring flowers. Its favorite haunt is rocky stream bluffs, associated with hepatica, crested-iris and giant-chickweed. Its rosette of leaves resembles those of the alum-roots. Flowers are borne in showy racemes. The fruits are interesting in that they are composed of 2 unequal valves which are not fused. It is a common plant in the Piedmont and mountains and is now being successfully transplanted to suitable locations in gardens. Ga. to e. Can.

Tiarella cordifolia · white

BROOK-SAXIFRAGE (*Boykinia aconitifolia*) As the
name suggests, this plant dwells in the freshness
and peacefulness of mountain brooks. The generic name comes from an early amateur botanist of Ga., Dr. Samuel Boykin. It has a leafy
stem up to 2″ tall, covered with glandular hairs,
and leaves mostly basal with 5-7 deeply-lobed
blades. Relatively small white flowers are borne
in a flat-topped inflorescence called a cyme. It
is to be looked for especially in the misty atmosphere along rapids and water falls. Blue Ridge
and Appalachians, Ga. to Tenn., N.C. and W.Va.

Boykinia aconitifolia   white

EARLY-SAXIFRAGE, ST. PETER'S-CABBAGE (*Saxifraga
virginiensis*) This is the commonest and most
widespread of the saxifrages, growing on stream
banks, shallow soil surrounding rock exposures,
and dry soil of poor drainage. It has a rosette
of leaves with ovate to elliptic blades and a stout,
naked flowering stalk with a flat-topped cyme of
white flowers with yellow stamens. When the
early-saxifrage is in bloom, it is a sign that spring
is well on its way. It is especially attractive in
small colonies and should be in everybody's wild
flower garden. Ga. to Ala. and n. to e. Can.

Saxifraga virginiensis - white

GRASS-OF-PARNASSUS FAMILY (Parnassiaceae)

CAROLINA GRASS-OF-PARNASSUS (*Parnassia caroliniana*) These plants with interesting flowers and
distinctive leaves were named for Mount Parnassus in Greece because one of them was thought
by the early botanists to be the grass of Parnassus
described in the *Materia Medica* of Dioscorides.
They are smooth perennials with basal leaves
and naked one-flowered stalks. The flowers are
interesting with their green-striped, white petals.
This species is distinguished by its clawless petals, obtuse anthers, and ovate cordate leaf blades.
Blooms in late summer. Swamps and flatwoods,
Coastal Plain, Fla. to N.C.

[45]

Parnassia glauca 1 - white

Hydrangea quercifolia - white

HYDRANGEA FAMILY (Hydrangeaceae)

OAK-LEAVED HYDRANGEA (*Hydrangea quercifolia*) This hydrangea was discovered and named by one of our earliest botanical explorers, William Bartram (1739-1815) of Philadelphia, in his adventurous travels throughout the Southeast. Of the numerous plants he discovered, surprisingly few were described and named by him. The oak-leaved hydrangea is a shrub with reddish-hairy twigs and large, deeply lobed leaves. Large panicles are composed of numerous small, white, fertile flowers and several sterile flowers with large petals, characteristic of other Hydrangeas and some Viburnums. The flowers are delightfully fragrant. It is native on river banks and bluffs of the Coastal Plain, extending into the Piedmont, n. Fla. to Miss. and Ga. It is, however, cultivated far beyond its natural range.

Itea virginica - white

VIRGINIA-WILLOW, SWEET-SPIRE (*Itea virginica*) This is a delightful shrub which one encounters in thickets on the margins of streams and swamps in various districts of our area. The conspicuous white racemic spikes appear during late April and continue into June. No other shrub is likely to be confused with this one in such habitats. It has horticultural possibilities in moist shaded situations. It keeps well when cut and makes a showy display in arrangements. Fla. to La. and n. to N.J.

white

Grossularia curvata   Rubus species

GOOSEBERRY FAMILY (Grossulariaceae)

GRANITE-GOOSEBERRY (*Grossularia curvata*) An attractive shrub, diffusely branched with gracefully arching, shining, reddish-purple branches with slender spines and 3-lobed, toothed leaves. It is a profuse bloomer with milky-white flowers with reflexed petals and long, exerted stamens. The globular berries, tipped with the persistent stamens dangle down the fruiting branches. It is limited to rocky slopes of the Piedmont of Ala. and Ga. It has wonderful horticultural possibilities as an ornamental, and the fruits may be used for jellies and preserves.

[46]

ROSE FAMILY (Rosaceae)

The rose family is noted for its flowers and fruits, and nearly all of our introduced genera are represented by relatives in our native flora.

BOWMAN'S-ROOT, FALSE-IPECAC (*Gillenia trifoliata*) This is an attractive perennial herb growing up to 5′ tall, with 3-foliate compound leaves and relatively large flowers with long narrow petals and reddish calyx tubes. It occurs in open woods in clayey soil, which often seems inhospitable to most plants of the Piedmont and mountains in the South. Ga. n. to e. Can. and w. to Mich. and Mo. Another species (*G. stipulata*), with large, deeply cut stipules and more deeply dissected leaves, is less common in this area. Both have distinctive ornamental values.

SNOW-WREATH (*Neviusia alabamensis*) This attractive shrub is well-named for it is rare outside Ala. It is now becoming widely cultivated not only in its native state but beyond. While it has no petals, its many whitish stamens and corolloid toothed sepals make it not only a conspicuous but distinctive flowering shrub. Leaf-blades are simple, ovate, and vary so much in size that it sometimes appears to bear two kinds of leaves. It grows to a height of 2′-5′, and is especially suitable for corners and park planting.

FLOWERING RASPBERRY, RASPBERRY ROSE, THIMBLE-BERRY (*Rubus odoratus*) The rose-purple blossoms of the flowering-raspberry attract attention even at some distance from the trails and highways of the s. Appalachian Mountains during the summer months. Since it commonly grows in thickets along the mountain streams, it is not always easily accessible for close inspection; but when this is attained, it repays the effort. It bears large palmately lobed leaves on clammy-pubescent stems and flowers in corymbose or racemose inflorescences. The fruits which resemble raspberries are very fragrant as if dipped in perfume, hence the specific name *odoratus*. Rocky banks and woods, Ga. n. to e. Can.

[47]

Gillenia trifoliata – white or pinkish

berry black × white
Rubus species–Neviusia alabamensis

Rubus odoratus – rose-purple

GEORGIA AND ALABAMA SPRING BOUQUET

GEORGIA AND ALABAMA SPRING BOUQUET

These flowers are found throughout the Piedmont and southern mountains. All last well and are suitable for moderate picking and for planting in the wild-flower garden.

Carolina Yellow Jessamine

Nun's-Hood Orchid

Dwarf Coral Bauhinia

CAROLINA YELLOW JESSAMINE (*Gelsemium sempervirens*)   Logania Family

In the early spring when much of the vegetation in the Coastal Plain is still dormant, the flower buds of the yellow jessamine commence to open from the ground up to the tallest trees on which it climbs in the pinelands, swamps, and sandhills. Later some of the pines turn yellow with the flowers of this high-climbing woody vine. It has opposite lanceolate, evergreen leaves. The flowers are borne in short axillary clusters and are delightfully fragrant. But the trumpet-shaped corollas soon drop after opening. It is easily transplanted to fences and trellises where it is a showy ornamental and an early harbinger of spring. It ranges from Fla. to Tex., Ark., and s.e. Va.

NUN'S-HOOD ORCHID (*Phaius grandifolius*)   Orchid Family

A glorious, showy, ground orchid, one of the easiest orchids grown by the unskilled gardener. Blooms in early spring; its 3'-4' spike, of from 15-20 multicolored, star-shaped blossoms, is borne above the long plaited leaves resembling those of young palms. Three sepals and 2 petals, white outside and red-brown inside, make a striking background for the third petal, which in characteristic orchid fashion has become modified into an exquisitely veined, tubular, spurred lip, dexterously folded around the reproduction organs. Rosy-purple, traced with copper, it turns to a rich copper-brown margined with yellow. Pot plant, housed during frost. Well-drained, rich soil, liquid manure, half shade.

DWARF CORAL BAUHINIA (*Bauhinia Galpinii*)   Pea Family

Bauhinia Galpinii is a graceful shrub with gorgeous salmon blossoms somewhat resembling nasturtiums. Though harder to grow and more susceptible to frost than *B. variegata,* its exquisite flowers, coming twice a year, and its striking dwarf habit, warrant its wider use. In India and Malaya its inner bark is used for rope and cordage. Layering difficult. Soak seed in water; protect from frost; prune.

Rosa carolina — pink

PASTURE-, LOW-, or CAROLINA-ROSE (*Rosa carolina*) The charm of wild roses cannot be surpassed by the most gorgeous of those we have cultivated. Although nowhere especially abundant, the Carolina-rose is widely distributed in e. N.A. It grows mainly in dry, rocky soil and is frequently seen in May and June in roadside clearings and open woods. Flowers are often solitary with pink petals and glandular, hispid pedicels; sepals, glandular on the back and matted-woolly within. It differs from the Virginia-rose (*Rosa virginiana*) in having more coarsely serrate leaves and more slender infrastipular prickles. N. Fla., Kans., Ky., N.C. and naturalized farther n.e.

Rosa palustris — pink

SWAMP-ROSE (*Rosa palustris*) Although the flowers of the swamp-rose are similar in color and form to the above, it cannot be confused with any other rose in the habitat in which it grows. It is generally robust with finely-toothed leaves and flowers more often clustered. The prickles are straight but stout. It is a common plant in marshes and borders of ponds and streams, in various districts from Fla. to Miss. and n. to the northern states and e. Can.

Rosa laevigata — white

CHEROKEE-ROSE (*Rosa laevigata*) An extensively climbing rose, introduced from China and so well naturalized in some of the southern states as to appear native. Petals are commonly white but occasionally pinkish. It is an impressive sight in the early spring to see festoons of the Cherokee-rose high up in the branches of the giant trees of the outer Coastal Plain swamps. It is often cultivated but because of its rank growth must be kept under control. Coastal Plain and adjacent areas, Fla. to Tex. and N.C.

[50]

Southern Crab-Apple (*Malus angustifolia*) Nothing is more delightful than the fragrance of the southern crab-apple when in full bloom with its numerous relatively small pinkish flowers. The leaves are narrow and blunt. It is a desirable tree to have in the neighborhood of your home. Unfortunately, it is often severely attacked by apple- or cedar-rust which blemishes its foliage. Found in thickets and edges of woods in various provinces, but more commonly in the Coastal Plain, from w. Fla. to La. and n. to Ill. and Va. The northern crab (*Malus coronaria*), (color page facing 32) with broad pointed leaves, extends into the southern Appalachian region.

Malus angustifolia — pink

Shrubby Service-Berry, Shad-Bush, June-Berry (*Amelanchier*) In the spring when the hardwood trees are beginning to leaf out, the white flowered racemes of the shad-bush may be seen in various situations throughout the southern states. The 5 white petals are long and narrow, with many stamens at base inserted on the rim of the cup-shaped calyx tube. The fruit is a berry which varies in taste in the different species. Small lists 6 species in the range of his *Manual of the Southeastern Flora,* two of which are tree size and occur in the mountains. The shrubby forms are more typical of the Coastal Plain.

Amelanchier canadensis — white

Parsley-Haw (*Crataegus Marshallii*) Hawthorns are notorious for being difficult to identify and for their vicious spines. But parsley-haw is an exception in being easily recognized by its characteristically lobed leaves, by its scarlet fruits in the fall, and by scarcity of spines. In spring it is covered with corymbs of white flowers with pink anthers. Low grounds, Coastal Plain, Fla. to Tex. and Va. Wedge-leaved haw (*Crataegus spathulata*) is similar to the above in many respects except for the shape of its leaves. It is also a red-fruited species. Coastal Plain, Fla. to Tex. and Va. Hawthorns are highly decorative with their flowers in spring and fruits in autumn.

[51]

Crataegus Marshallii C.spathulata

Chrysobalanus oblongifolius C. Icaco

white

cream   white

Prunus caroliniana Prunus umbellata

purplish,
to
Calycanthus floridus—maroon

COCO-PLUMS and GOPHER-APPLES (*Chrysobalanus*) These shrubs are closely related to peaches, almonds and apricots and occur mainly in s. Fla. although some of them extend into Ga. Coco-plum (*Chrysobalanus Icaco*) may attain small tree size and when growing on beaches has radially creeping branches. Flowers are white and the drupes yellow, red, or purple. They are edible when used as preserves. When planted as an ornamental it thrives away from the coast. Gopher-apple or ground oak (*Chrysobalanus oblongifolius*) is a relatively low plant with underground stems, growing in habitats similar to coco-plum. Its leaves resemble certain narrow-leaved oaks. Flowers are white and the oblong fruit, ivory-white, tinged with purple or red and fragrant. Fla. to Miss. and Ga.

HOG-PLUM, SLOE, BLACK-SLOE (*Prunus umbellata*) A small tree up to 24' with umbels of white 5-petalled flowers. Its reddish to purple fruits are a favorite food for birds. Decorative and highly prized for ornamental planting, blooming from late January to early March. Dry, often sandy, open woods, along streams and hedgerows. Coastal Plain, Fla. to La. and S.C.

CAROLINA CHERRY-LAUREL, MOCK-ORANGE (*Prunus caroliniana*) An evergreen, small tree with finely serrate, narrowly elliptic shining leaves. Small white flowers are borne in dense racemes in late winter or early spring. In woods usually on river banks and in hammocks, native to Coastal Plain from Fla. to Tex. and N.C. but widely cultivated and escaping.

SWEET-SHRUB FAMILY (Calycanthaceae)

SWEET-SHRUB, STRAWBERRY-SHRUB (*Calycanthus floridus*) Sweet-shrubs or strawberry-bushes are among our most delightful native plants. Their strawberry scent is sometimes detected at considerable distances. They vary considerably in fragrance and seem to vary in different years depending on the season. The flowers have numerous dull-purple sepals and petals and numerous stamens. The fruits, or "seeds," are borne in an urn-shaped hypanthium similar to that of a rose. A western species (*C. occidentalis*) is sometimes planted in the East.

[52]

PEA FAMILY (Leguminosae)

CAT'S-CLAW, BLACK-BEAD (*Pithecellobium Unguis-Cati*) The generic name of this shrub or small tree means "monkey earring" in allusion to its long, spirally-twisted seed pods; the specific name comes from its vicious spines resembling cat claws which may often be absent entirely. It is distinguished from other members of the mimosa tribe by its few leaflets (usually 4), numerous stamens, and the thick, leathery, twisted pods. The showiness of the flowers consists of the numerous, long, brush-like stamens which range from cream to pink. It blooms in late spring. Hammocks of s. Fla.

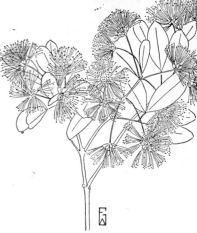

Pithecellobium Unguis-Cati -mauve

LEAD-TREE (*Leucaena glauca*) A shrub or small tree, usually about 8'-15' but may attain a height of 30', with bipinnate leaves of numerous, small leaflets. The flowers are arranged in a spherical head similar to many others of this tribe, and the stamens are white to pinkish. The pods appear in green to brownish clumps and hang on the tree at all seasons of the year. In hammocks and cleared ground, Fla. to Tex.

SWEET-ACACIA, POPINACK, YELLOW-OPOPANAX (*Vachellia Farnesiana*) A stout shrub or small tree which has immigrated from its original home in the American tropics and escaped from many southern gardens and become widely naturalized in hammocks, pinelands, and waste places in the Coastal Plain. It has bipinnate leaves and small, yellow balls of flowers and stout pods. Fla. to Tex.

white or pink / *yellow*

Leucaena glauca-Vachellia Farnesiana

SENSITIVE-BRIARS, SENSITIVE-PLANTS (*Schrankia*) These interesting plants are familiar to anyone who has tramped dry, open woods or pinelands. If not attracted by the pink or rose-purple puffy balls of flowers, he undoubtedly noticed the spectacular closing of the delicate bipinnate leaves when disturbed by touch and even by a sharp sound. The leaves are covered with sharp hooked spines, hence the name "briars." Five species are listed for the South in Small's *Manual of the Southeastern Flora*. Of these, *S. floridana* is the most southern, being found only in pinelands of Fla. *S. Chapmanii* ranges from Fla. to Tex. and N.C. and *S. Nuttallii* and *S. microphylla* have wider distributions.

[53]

fruit / *leafletX*

Schrankia floridana Mimosa strigillosa

RED-BUD, JUDAS-TREE (*Cercis canadensis*) When the red-bud begins to bloom and the redbird sings, "Boys, boys, do, do," spring has come to the Southland. It blooms about two weeks before the flowering dogwood but they overlap in their flowering periods at least for a few days. It is one of the few members of this family with simple leaves which are heart-shaped. The pods are small and flat. When isolated in landscaping, it flowers more profusely than when crowded by other trees as is so often the case in nature. It grows in various situations, often in dry soil in various provinces from Fla. to Tex., n. to Neb., Ohio, and Ont.

Cercis canadensis - pink
seed pod

yellow   yellow
Cassia ligustrina - C marilándica

nectar gland
Cassia brachiata - yellow

WILD-SENNAS, CASSIAS (*Cassia*) Our native Cassias are mostly annual or perennial herbs with pinnately compound leaves and yellow flowers. The petals are nearly equal in shape but the stamens are different. The lower anthers are large and functional with a pore at the top while the upper are short and sterile. The flowers are relatively large and attractive, blooming in late summer and in some localities so abundant as to color open areas a solid yellow. The leaflets fold up at night. *C. ligustrina*, closely related to coffee-weed or nigger-coffee (*C. occidentalis*), is large and partly woody with pointed leaflets. It grows in hammocks of peninsular Fla. and blooms the year around. Wild-senna (*C. marilandica*) may grow up to 5′ tall with 12-20 elliptic leaflets. It grows on moist stream banks and alluvial soil in various districts from N.C., Tenn. to Mass. Sensitive-pea (*C. brachiata*) is a coarse annual, 1′-4′ tall, with reddish stems and 8-18 tiny leaflets with small, saucer-shaped nectar glands at base. Common in sandy soil, especially in the Coastal Plain, Fla. to Ala. The Seminoles use a cold water extract of this for cases of nausea. The partridge-pea or golden-cassia (*C. fasciculata*) is the most frequent and widely distributed species in e. U.S. and is especially abundant in some of the open, sandy areas in the outer Coastal Plain.

AARON'S-ROD (*Thermopsis caroliniana*) A perennial herb from 1'-3' tall with spreading branches and 3-foliate leaves and racemic-spikes of relatively large crowded, yellow flowers. It can be distinguished from yellow-flowered indigos by the woolly, hairy flowering stalk and woolly legumes. This plant grows in dry soil from sandy pinelands of the Coastal Plain to clayey soils of open woods in the Blue Ridge Mountains. It blooms in the spring and the showy yellow spikes always attract attention. Under cultivation it grows much larger than in the wild state and bears much longer flowering stalks with larger flowers. Ga. to Tenn. and N.C.

Thermopsis caroliniana — yellow

FALSE- or WILD-INDIGOES (*Baptisia*) These are also perennial herbs with 3-foliate leaves and the majority of them have yellow flowers. The fruits are characteristically short-stalked and swollen. Eighteen species are listed in Small's *Manual of the Southeastern Flora*. Of the yellow-flowered species, *B. tinctoria* is perhaps the most frequent and widespread. It has scattered flowers in the axils of the upper leaves. *B. lanceolata* has scattered flowers in a terminal, naked raceme and grows in dry pinelands from Ala. to N.C. *B. leucantha* is one of the white-flowered species easily recognized by its drooping fruits. Fla. to Tex n. to e. Can. Blue-indigo (*B. australis*) is often cultivated and escapes to roadsides and waste ground. Ga. to Tenn. and Pa.

Baptisia leucantha – B. lanceolata — white, yellow

LUPINES (*Lupinus*) The lupines are showy and attractive with their gorgeous spikes of mostly blue flowers, and some exotics are grown in our gardens. The southern lupines grow mostly in sandy soil of the Coastal Plain and are, therefore, known as sandhill-lupines. Nuttall's-lupine (*L. Nuttallii*) is our only one with digitate leaves. It was named for an early botanist of Philadelphia, Thomas Nuttall (1786-1859), who wrote the first comprehensive flora of this country. It is most successfully grown from seed. Two sandhill lupines with simple leaves are *L. diffusus* with blue flowers which ranges from Fla. to Miss. and N.C., and lady-lupine (*L. villosus*), a remarkable species with deep-lilac to purple corollas with a red spot in the center of the standard. Blue-bonnet (*L. texensis*) is the state flower of Texas.

[55]

Lupinus Nuttallii — lavender, *L. diffusus* — blue, *L. villosus* — red

RABBIT-FOOT or OLD-FIELD CLOVER (*Trifolium arvense*) Aside from their value as soil improvers and for fodder, most species of clover are highly ornamental. Nothing is more beautiful than a field of crimson clover in full bloom. The same may be said also of a field of red- or purple-clover which, however, is seldom seen in the South. Rabbit-foot clover is of artistic value because of its grayish-hairy spikes. It is not utilized in agriculture in this country but grows mostly as a harmless weed on roadsides and in waste places. It has a distinctive place in flower arrangements which has so far not been fully appreciated. Naturalized from Eurasia and widely distributed in e. N.A.

iridescent blueish To purple
Trifolium arvense. Psoralea psoralioides

SAMPSON'S-SNAKEROOT, CONGO-ROOT (*Psoralea psoralioides*) The Psoraleas are perennial herbs growing from deep roots or rootstocks in usually acid, sandy soil. The bluish flowers are borne in spikes or spike-like racemes. The campanulate calyx is usually glandular-dotted. The number of leaflets vary but there are rarely more than 3. Congo-root ranges in the Coastal Plain from Fla. to Tex. n. to Ind. and Va. *P. virgata* is unique in having mostly one leaflet. It is more southern, being restricted to Fla. and Ga. *P. subacaulis* is a curious plant resembling more a lupine because of its digitate compound leaves. It is restricted to the Cedar Glades of Tenn. It is sometimes placed in another genus (*Pediomelum*) and perhaps rightly so.

9 stamens X
Psoralea subacaulis - blue x purple

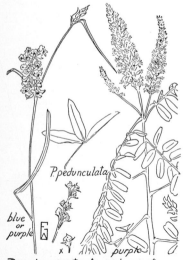

INDIGO-BUSH, PLUME-LOCUST, LEAD-PLANT (*Amorpha fruticosa*) These shrubby, leguminous plants have branched, racemic-spiked inflorescences which range in color from light-blue to purple. The small flowers are interesting in that they have only one petal (the standard), hence the name *Amorpha* meaning "off form." The compound leaves are composed of several leaflets marked with minute dots. Besides this species, there are some 14 others in the South which are best represented in the Coastal Plain. Many of these are so hairy as to give them a grayish appearance. Fla. to La. and n. into northern states.

[56]

P. pedunculata
blue or purple
Psoralea virgata Amorpha fruticosa
purple

Spring Wild Flowers
of the Carolina Mountains

Linaria canadensis

Specularia perfoliata

Zizia aurea

Penstemon canesens

Barbarea
verna

Tephrosia virginiana

Geranium
maculatum

Zephyranthes
atamasco

Coreopsis grandifolia

Houstonia
caerulea

Viola pedata

Chrysogonum virginianum

Vicia angustifolia

Vicia grandiflora

Oxalis violacea Viola canadensis

Early Summer Flowers
of the Carolina Mountains

Stenanthium gramineum *Euphorbia corollata*

Verbascum Thapsus *Daucus carota*

Spiraea tomentosa *Linaria vulgaris*

Centaurea maculosa

Hypericum perforatum

Dicentra eximia

Asclepias incarnata var *pulchra*

Malva moschata *Aconitum uncinatum*

Monarda fistulosa

Rhododendron (Azalea) arborescens

Phlox glaberrima *Rudbeckia hirta*

Vaccinium pallidum *Prunella vulgaris*

Asclepias tuberosa

Midsummer Flowers
of the Carolina Mountains

Campanula divaricata

Thalictrum revolutum

Clethra
acuminata

Polygonum orientale

Saponaria officinalis   Campanula americana

Gerardia virginica

Scuttelaria incana

Echium vulgare   Monarda didyma

Clethra leaf

Sabatia angularis

Lobelia cardinalis

Verbascum Blattaria   Clematis virginiana

Fall Flowers
of the Piedmont and Coastal Plain

Cimicifuga racemosa   *Silphium*
Psoralea psoralioides   *compositum*

Chrysanthemum
leucanthemum

Coreopsis verticillata

Phlox maculata

Echinacea laevigata

Coreopsis major

Rosa palustris

Stokesia laevis

Polygala mariana

Helianthus divaricatus

Oenothera fruticosa

Cichorium Intybus

Clitoria mariana

RATTLE-BOXES, RATTLE-BELLS (*Crotalaria*) These plants are called rattle-boxes because when the seeds are mature, they come loose and rattle within the dry, inflated pod before it splits open. The flowers are yellow in our native species. Seaside rattle-box (*C. maritima*) is found in various situations in the Coastal Plain but is most frequent near the coast associated with coastal plants. The lower leaves are orbicular and consist of only one blade. An infusion from the pods is used by the Seminoles as a cure for sore throat. Fla. to La. and N.C. Pursh's rattle-box (*C. Purshii*) has narrow, linear to wedge-shaped leaves on erect or clustered stems. Pinelands, prairies, sand dunes, and grassy woods, Coastal Plain, Fla. to Tex. and Va.

yellow
Crotalaria Purshii - C. maritima

PURPLE-TASSEL, PURPLE PRAIRIE-CLOVER (*Petalostemum purpureum*) This is the showiest of the prairie-clovers, herbs with few to many leaflets and dense spikes of purple flowers with short, broad standards. Some, however, have white flowers. Purple-tassel has violet or rose-purple corollas and heads the species which are suitable for gardens. Widely distributed in e. and c. U.S.

rose-purple

Petalostemum purpureum - Indigofera caroliniana

CAROLINA INDIGO-PLANT (*Indigofera caroliniana*) A perennial with slender branches, 2'-6' tall, and pinnately compound leaves of 9-15 elliptic leaflets. The small flowers are borne in slender racemes in the upper leaf axils, varying in color from yellow-brown to peach or dull red. The distinguishing characteristic is the scimitar-shape keel petal with auricle-lobed wings at base. The pod is mostly 2-seeded, rarely 3. Dry pinelands and scrub of the lower Coastal Plain from Fla. to La. and N.C.

SUMMER-FAREWELL (*Kuhnistera pinnata*) This is an unusual-looking herbaceous plant from 12"-50" tall with 3-7 linear-clavate leaflets and dense heads of white flowers surrounded with reddish bracts. A late summer bloomer, its fruiting heads persist into the winter and are gray because of the copiously hairy calyx lobes. In this condition, it makes beautiful dried bouquets. Dry pinelands and sandhills, Coastal Plain, Fla. to Miss. and N.C.

Kuhnistera pinnata - white

AMERICAN WISTERIA (*Wisteria frutescens*) Anyone familiar with the commonly cultivated Japanese- or Chinese-wisteria can easily recognize this native growing on banks of creeks and rivers and borders of swampy woods in the Coastal Plain. Like most Wisterias, it has an extensively twining stem. Occasionally planted on fences and trellises in the localities where it grows naturally. Small says, "The contrast in color of the early foliage and the purple flower-clusters is a pigmentary perfection even Nature rarely attains." Blooms in late spring. Coastal Plain and rarely adjacent provinces, Fla. to Ala. and Va.

Wisteria frutescens - purple

ROSE-ACACIA, MOSS-LOCUST, BRISTLY-LOCUST (*Robinia hispida*) A shrub up to 6' tall with pinnately-compound leaves of 7-15 orbicular to ovate leaflets and racemes of 3-5 pink to pale-purple, sweetpea-like flowers. The stems as well as the fruits are densely hispid with reddish bristles. Because of its escape from extensive cultivation, it is now difficult to tell just what its natural range was originally. It occurs now naturally on dry slopes in various districts, except in the Coastal Plain, from Ga. to Ala. and n. to Ill. and Va. It is highly variable and is known to hybridize with black-locust. Several other species of locusts occur in the South.

Robinia hispida - pink

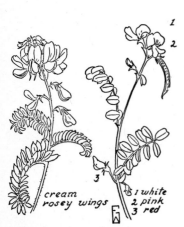

SHOE-STRINGS, CAT-GUT, DOLLY-VARDEN (*Tephrosia virginiana*) A loosely-hairy perennial with many ascending stems and pinnately compound leaves of 11-21 elliptic leaflets. The pea-like flowers are medium-sized with a pale-yellow standard and pink or purplish wings and keel. Although the flowers are individually showy, they are partly hidden by the dense foliage. It is a common flower of usually dry acid or sandy soil and, although most common in the Coastal Plain, extends into other areas, blooming in late spring. Fla. to La. and n. to e. Can. Other species of *Tephrosia* are mostly low, sparsely-branched perennials with white to cream flowers which turn red with age.

cream
rosey wings

1 white
2 pink
3 red

[58]

Tephrosia ((racca)virginiana - T. ambigua

JAMAICA-DOGWOOD (*Ichthyomethia piscipula*) The common name of this deciduous small tree is misleading, for the upright panicles of white, pea-like blossoms resemble the northern locust-tree. They are delicately streaked with lavender or red giving an opalescent, mauve effect. It has a soft gray-green foliage of 7-9 pointed leaflets which do not unfold in the spring until the panicles are in full bloom. Bees are fond of the flowers, and the hard, durable wood is valuable for boat-building. The bark of the roots and stems was formerly dried, powdered, and used to stupefy fish. Hammocks of s. coastal regions of Fla. and the Keys.

white streaked purple

Ichthyomethia piscipula

CORAL-BEAN, CHEROKEE-BEAN, RED-CARDINAL (*Erythrina arborea*) In habit this varies from a shrub to a small tree or vine, with compound 3-foliate leaves of distinctive blunt, hastate leaflets. Racemes generally few-flowered of narrow, elongate perianths with a standard up to 15″ long, red, scarlet, or crimson and more or less drooping. The drooping pods are contracted between the bright-red seeds, which are often gathered and used for beads. Hammocks of southern peninsular Fla. and the Keys. A closely related perennial herb called cardinal-spear (*E. herbacea*) is very similar in its leaf, floral, and fruit characters. In pinelands, hammocks, thickets, and scrub of the Coastal Plain from Fla. to Tex. and N.C.

Erythrina arborea-red

KUDZU VINE (*Pueraria lobata*) After being introduced from eastern Asia a number of years ago, this robust woody vine has rapidly invaded many areas of the South covering road-banks, fence-rows, and even woods to such an extent that it is becoming a pest as has the Japanese honeysuckle. It has been cultivated to some extent for soil improvement and fodder but, after it has become well established, it becomes difficult to control or eradicate. Covering many bare areas, it undoubtedly does some good in preventing soil erosion. The dull-reddish flowers which appear in late summer and fall are borne in drooping racemes. Widely distributed in the South.

Pueraria lobata - purplish

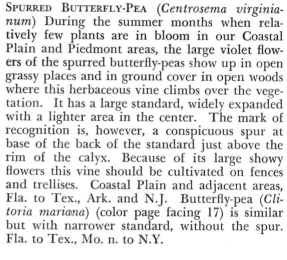

Lilac to blue

Clitoria mariana · Centrosema virginianum

SPURRED BUTTERFLY-PEA (*Centrosema virginianum*) During the summer months when relatively few plants are in bloom in our Coastal Plain and Piedmont areas, the large violet flowers of the spurred butterfly-peas show up in open grassy places and in ground cover in open woods where this herbaceous vine climbs over the vegetation. It has a large standard, widely expanded with a lighter area in the center. The mark of recognition is, however, a conspicuous spur at base of the back of the standard just above the rim of the calyx. Because of its large showy flowers this vine should be cultivated on fences and trellises. Coastal Plain and adjacent areas, Fla. to Tex., Ark. and N.J. Butterfly-pea (*Clitoria mariana*) (color page facing 17) is similar but with narrower standard, without the spur. Fla. to Tex., Mo. n. to N.Y.

Cream-purple tip

Strophostyles umbellata

white, pink, red

Galactia Elliottii

SAND-BEANS, TRAILING-BEANS, WILD-BEANS (*Strophostyles umbellata*) Trailing- or sand-beans differ from true beans in their more common trailing habit, in the long naked flower-stalk and in the incurved but not spirally twisted keel. *S. umbellata* is the most common and widespread species growing in dry sandy fields and woods. The pink or pale-purple corolla often fades to yellow. Coastal Plain and lower Piedmont, Fla. to La., Mo., and N.Y. Elliott's milk-pea (*Galactia Elliottii*) is a unique species of *Galactia* in having 7-9 evergreen leaflets and white flowers. Fla. to Tenn. and N.C. *Phaseolus sinuatus* is a true perennial bean with purple flowers. A trailing plant of dry pinelands and hammocks. Fla. to Miss. and N.C.

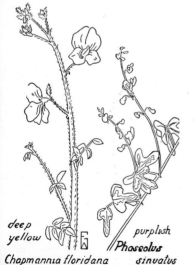

deep yellow

Chapmannia floridana

purplish

Phaseolus sinuatus

ALICIA, CHAPMAN'S PEA (*Chapmannia floridana*) (color page facing 17) Named by John Torrey and Asa Gray in honor of one of our most outstanding southern botanists, Alvin Wentworth Chapman (1809-1899), author of *Flora of the Southern United States*. An erect, hairy, perennial herb with 5-7 pinnately compound leaves and two kinds of flowers, one perfect, or bisexual, but not fruit-producing, the other pistillate and setting fruit. Corolla yellow, falling soon after opening, with a two-lipped calyx and a 1-4 jointed fruit. The only species of this genus (monotypic). Grows in dry pinewoods, peninsular Fla.

[60]

BAY-BEAN, JUNE-BEAN (*Canavalia lineata*) One of the first indications of land vegetation seen by Columbus on his famous voyage to the West Indies was floating seeds of the bay-bean. This extensively trailing vine with its 3-foliate large, fleshy, suborbicular to oval leaflets grows in the coastal sand dunes of the Gulf states from Fla. to Tex. and the W.I. The flowers are pink or rose-purple and the pods are from 3½"-4" long. The related horse-bean (*C. gladiata*) has a white or purple corolla sometimes tinged with yellow and a much larger pod. Naturalized from the W.I., it occupies waste places and cultivated ground. Fla.

Canavalia lineata - rose-purple or pink

VETCHES, TARES, or PEA-VINES (*Vicia*) (color page facing 17) Vetches are like the sweet- and garden-peas in being provided with tendrils at the end of the leaves which support them in climbing on artificial or natural supports. Several of our annual vetches have been introduced intentionally or inadvertently and some of these are used as winter cover-crops and some for ornamental purposes. Canada-pea or cow-vetch (*V. Cracca*) is an introduced perennial with blue to purple flowers in a raceme, naturalized in fields and waste places. Hairy- or winter-vetch (*V. villosa*) is an introduced annual, or biennial, spreading and hairy, with long, congested racemes with purplish (rarely white) flowers. Carolina- or wood-vetch (*V. caroliniana*) is a beautiful, native perennial with racemes of white flowers except for the purple-tipped keel. Various localities, Ga. to Kans. and n. to w. N.Y.

white keel blue tipped

bluish

Vicia caroliniana - *V. acutifolia*

CRAB'S-EYE, INDIAN-LICORICE, ROSARY-PEA (*Abrus precatorius*) The outstanding characteristic of this extensively climbing vine is the scarlet seeds with a black base which are used for beads and novelties wherever it grows. It is native to Europe but has become naturalized in peninsular Fla. and various tropical regions of the western hemisphere. The flowers are mostly pink, the keel petals are longer than the wings, and the seed pods are clustered. The bright-colored seeds which are said to contain an irritant poison were used at one time as a standard of weight in India. Woods, thickets, and roadsides.

[61]

Abrus precatorius - pinkish - seeds red

D. Chapmania

bluish-
to purple

purplish

Desmodium triflorum-D. rhombifolium

BEGGAR'S-TICKS, BEGGAR-LOUSE, TICK-TREFOILS (*Desmodium*) Among the assortment of various fruits that stick to your clothing as you wander through the woods on autumn days, the most common are the joints of the fruits (loments) of *Desmodium*. The easiest way to remove them is with a boll of cotton. About 30 species occur in the South, and they seem to be everywhere. They are mostly herbaceous perennials with trifoliate leaves which in some species are basal and in others occur on the stem. The 3 illustrated are typically southern. *D. triflorum,* which is sometimes placed in the genus *Sagota* is tropical. It resembles a lespedeza and is used as a lawn cover.

Oxalis violacea -violet

WOOD-SORREL FAMILY (Oxalidaceae)

WOOD-SORRELS, SOUR-GRASSES (*Oxalis*) The outstanding vegetative characteristic of the wood-sorrels is the compound leaf blade with 3 (or 4) heart-shaped leaflets attached at the tip (obcordate), and, second, the sour taste, due to the presence of oxalic acid as in sour-wood. Several of the southern species are yellow, some of which are adventive as weeds in our gardens, while others are introduced as ornamentals. In the damp woods of the southern mountain districts is found wood-shamrock (*O. montana*), with white to pink flowers, but the violet wood-sorrel (*O. violacea*), with rose-purple (rarely white) petals is more widespread and is often cultivated. Fla. to Tex. and n. to Me.

FLAX FAMILY (Linaceae)

YELLOW-FLAXES (*Linum*) The bark of a species of flax (*L. usitatissimum*) is the source of fiber for linen and paper, and the seed, the source of linseed oil. A field of cultivated flax with its delicate blue flowers is a beautiful sight in the early morning, for as the day advances they soon wither. Several introduced kinds are cultivated as ornamentals, especially in rock gardens. Most of our native species of flax have yellow flowers and are summer bloomers, or, in more southern areas, they bloom all year. They are slender herbs with relatively small, narrow leaves and are often seen in open, grassy places.

yellow

Linum (Cathartolinum) Carteri-L. floridanum

JEWEL-WEED FAMILY (Balsamiaceae)

Touch-Me-Not, Jewel-Weed (*Impatiens*) When you touch the mature fruits of the jewel-weed you are startled by the wormy feel of the elastically opening capsule. Hence the name, touch-me-not. The flowers with their nectar sac look like miniature, decorated pitchers with a recurved or bent spur at the base. Two native species are found in the South: spotted-snapweed (*I. capensis*), orange or reddish with black spots and long spur, and pale touch-me-not (*I. pallida*), yellow (rarely creamy-white), sparingly spotted and a short spur. Both are annuals, growing in seepages, edges of springs, rivers, and ponds. Piedmont and mountains.

-orange, black spots

Impatiens capensis

CALTROP FAMILY (Zygophyllaceae)

Puncture-Weed, Bur-Nut (*Tribulus cistoides*) This family includes the well-known lignum-vitae (*Guaiacum sanctum*), a shrub or tree which furnished a resinous heartwood famous for its hardness, weight, and strength. *Tribulus* means trouble, for the fruits of puncture-weed attach themselves by means of barbed spines to clothing and even to the skin. But the flowers are a beautiful bright-yellow. An annual herb, adventive and naturalized from tropical America. Hammocks and waste places, Coastal Plain, Fla. to Tex. and Ga.

Tribulus cistoides - yellow

MALPIGHIA FAMILY (Malpighiaceae)

Locust-Berry (*Byrsonima cuneata*) This is the only native representative in the South of this tropical and subtropical family. It is a shrub or tree with opposite, evergreen, lustrous leaves and racemes of white or pink flowers, fading yellow, with prominently stalked, kidney-shaped petals. The fruit is an attractive drupe about 1″ in diameter. Hammocks and pinelands, tropical Fla. Representatives of 3 related Caribbean and S.A. genera, the best known of which is Barbados-cherry (*Malpighia glabra*), are cultivated in the South for their profusion of unique, yellow or pink flowers.

[63]

pink

Byrsonima cuneata Locust berry

Blazing Star *Liatris spicata*   Hard-Head *Xyris pollescens*

Blue-Eyed Grass
Sisyrinchium graminoides

Thimbles-Yellow Milkwort
Polygala Rugelii

Orange Red Lily
Lilium Catesbae

Orange Fringed Orchid Iron weed
Blephariglottis ciliaris   *Vernonia*
Habenaria

Crotalaria
Crotalaria striata

Partridge Pea
Chamaecrista brachiata

Goldenrod   Rabbit Tobacco
Solidago Chapmanii *Pterocaulon undulatum*

Sunflower
Helianthella grandiflora

Sabbatia Sea Star   Golden Aster
Sabbatia Elliottii   *Chrysopsis*

Trilisa paniculata

Butterfly Weed
or
Blazing Star   Milkweed   Sneezewood
Helenium   Red Top Grass

Liatris pauciflora   *Asclepias Tuberosa*   *Tricholaena rosea*

FALL WILD FLOWERS
OF FLORIDA AND SOUTHERN COASTAL PLAIN

[64]

FALL WILD FLOWERS OF FLORIDA AND SOUTHERN COASTAL PLAIN

Catesby's lily and orange fringed orchid should be picked only where found growing abundantly, and then sparingly. Both are difficult to transplant to the wild-flower garden. All others may be picked freely. Liatris is easily grown and beautiful in the garden.

SUMMER AND FALL FLOWERING ORNAMENTAL SHRUBS AND VINES
Cultivated from coast to coast in warm areas.

Pink Bignonia
Pandorea ricasoliana

Pink Crape Myrtle
Lagerstroemia indica

White Crape Myrtle
Lagerstroemia indica alba

White Oleander
Nerium Oleander

leaf of
Pink Bignonia

Pink Acanthus
Jacobinia Carnea

African Daisy
Gerbera Jamesoni

Pink Oleander

Abelia
Abelia grandiflora

Snow Bush
Phyllanthus or *Breynia nivosa*

Azalea *Indica*
v. Duc Du Rohan

Double Pink Hibiscus

Pinwheel Flower
Tabernaemontana cashmere

Cestrum
Cestrum elegans

Coral Vine
Antigonon leptopus

SUMMER AND FALL
FLOWERING ORNAMENTAL SHRUBS AND VINES

Most of the shrubs and vines in this composite group will be found in line drawings with the text with the exception of the following: pinwheel-flower (*Ervatamia cashmere*) formerly known as *Tabernaemontana*, is one of the freest-flowering, glossy broad-leaved medium-sized evergreen shrubs for Zones 1-2; *Breynia nivosa*, better known as *Phyllanthus* is a decorative foliage shrub of the euphorbia family and has leaves variegated with pink and white or purple and white. It is used extensively in Zone 1 and parts of Zone 2. For cutting and arranging suggestions, see p. 149.

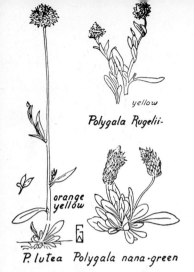

yellow

Polygala Rugelii

orange yellow

P. lutea   *Polygala nana-green*

yellow

purplish

Polygala cymosa   *P. grandiflora*

red bracts

flower green

Poinsettia pinetorum P. heterophylla

MILKWORT FAMILY (Polygalaceae)

MILKWORTS, CANDYWEEDS (*Polygala*) Milkworts do not give milk as their generic name implies (*Polygala* means much milk). The name was first applied to a shrubby relative which was in ancient times recommended for increase in lactation. Our native species are herbaceous annuals or perennials, with usually small, curiously-shaped flowers. The lower petal is called the "keel" which often has a hood or fringed "crest." Flowers are borne in spikes, racemes, or cymes and range in color from yellow, orange, shades of purple to greenish-white or white. The majority of the southern species occur in the Coastal Plain, the most conspicuous of which is the yellow-polygala (*P. lutea*), which is usually orange, with a short, condensed raceme. Among the smaller species are *P. cruciata* with leaves in whorls of 3 or 4 and usually purplish flowers; *P. Rugelii*, with yellow flowers, restricted to Fla.; and *P. nana*, called "bachelor's-button," with short greenish-yellow racemes. Two species with naked stems, basal rosettes of leaves and panicled yellow or greenish-yellow inflorescences are *P. cymosa* and *P. ramosa*. The largest-flowered species is *P. grandiflora* which, because of its large flowers, does not look like a *Polygala,* and is sometimes placed in another genus, *Asemeia.* This is an unusually attractive flower with its greenish-purple or bright-purple flowers. Fla. to Miss. and S.C.

SPURGE FAMILY (Euphorbiaceae)

This remarkable plant family is characterized by a flower group which resembles one flower and usually has a milky juice. It includes such ornamentals as the Christmas-flower (*Poinsettia*), snow-on-the-mountain, and the slipper-flower (*Pedilanthus*), and such important economic plants as the Pará rubber-tree of Brazil, castorbean, and the tung-oil tree.

POINSETTIAS (*Poinsettia*) A few relatives of the Christmas-flower (*Poinsettia pulcherrima*) are native in tropical Americas and some extend into our subtropical areas. This relationship is indicated by reddish areas at the base of the upper leaves. Five species occur in the South and a few extend as far north as Va. and Ill.

[66]

SPURGE-NETTLE, TREAD-SOFTLY (*Cnidoscolus sti-mulosus*) Most of us have received a rude intro-duction when we first picked this plant because of its stinging, nettle qualities. However, though the burning sensation may at first be alarming, it soon passes away with no ill effects. The attractive white "flowers" are male and fe-male. Dry, sandy soil, Coastal Plain, Fla. to Tex. and Va.

female and × *male flowers*
Cnidoscolus stimulosus *white*

FLOWERING-SPURGE (*Euphorbia corollata*) This is a common and widespread plant in which the white, petal-like bracts surround a group of small staminate flowers, in the center of which is a single pistillate flower. It is an odd and attrac-tive plant which grows in various situations and often in sterile soil. Widely distributed in e. N.A.

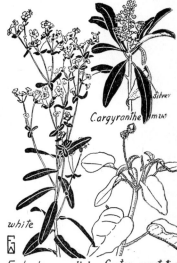

Silver
C. argyranthemus

white

Euphorbia corollata *Croton punctatus*

CROTONS (*Croton*) Several species of this genus occur in the South. The sexes of the flowers are separate on the same plant, and blooms are not showy. The showiness of Crotons is in their stems and foliage, and the latter is often covered with star-shaped hairs so thickly as to give the plant a gray or white, velvety appearance. Most of them are annual or biennial herbs but the Ala-bama-croton (*C. alabamensis*) is a shrub up to 9' tall, with large, evergreen leaves, covered be-neath with silvery, overlapping scales. It is cu-rious that it should be restricted to a limited area of limestone shales of Ala. Seaside-croton (*C. punctatus*) is coastal, growing on sand dunes. It is covered with stellate hairs. *C. argyranthemus* grows more inland. The under side of its leaves are covered with silvery scales as in the Alabama-croton. Fla. to Tex. and Ga.

[67]

× *pistillate* × *staminate flower*

Croton alabamensis

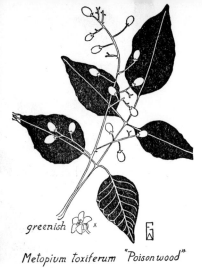

greenish   x

Metopium toxiferum "Poisonwood"

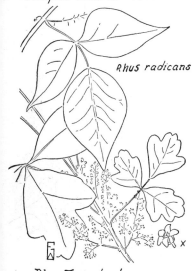

Rhus radicans

Rhus Toxicodendron

white

Cyrilla racemiflora-Cliftonia monophylla

SUMAC FAMILY (Anacardiaceae)

The sumac family includes the mango of tropical fame, the smoke-tree, poison-wood, poison-oak, and poison-ivy, and, of course the sumacs.

POISON-WOOD, CORAL-SUMAC (*Metopium toxiferum*) This shrub or small tree is to be avoided, for it causes some of the worst cases of dermatitis acquired from plants. It has evergreen, pinnately-compound leaves and dioecious, yellow-green, dark-lined flowers. The leathery 3-7 leaflets are 1½"-3¾" long. It frequents hammocks, pinelands, and coastal dunes, s. Fla., the Keys, and W.I.

POISON-IVY and POISON-OAK (*Rhus*) Shrub or vine with compound 3-foliate alternate leaves, causing more or less severe cases of dermatitis, apparently depending upon individual susceptibility. Considerable confusion exists as to which is which, and whether they are distinct or not. Poison-oak (*R. Toxicodendron*) is a low, sparingly branched, hairy shrub with thickish, prominently- and regularly-lobed leaflets, growing mostly in dry sandy or clayey soil. Poison-ivy (*R. radicans*) is a smooth vine which climbs by means of aerial rootlets, with thin, sparingly and irregularly lobed leaflets. Both have white to greenish, small flowers and waxy, berry-like (drupes) fruits. Widely distributed in e. U.S.

TITI FAMILY (Cyrillaceae)

BLACK TITI, LEATHERWOOD (*Cyrilla racemiflora*) Well-known for its brown-tinged, close-grained, hard and heavy heartwood, but famous for its honey, this small tree of swamps, pocosins, bays, and stream banks of the Coastal Plain is easily spotted at some distance by its sprays of slender racemes of white flowers blooming in late spring. S.e. Va., southward to Fla. and La.; also in W.I.

TITI, IRONWOOD, BUCKWHEAT-BRUSH (*Cliftonia monophylla*) Another source of excellent honey, this small evergreen tree is distinguished from *Cyrilla* by its terminal instead of lateral panicles and by its winged nut-like drupes. Only one species is known. Swamps and bays, n. Fla. to La. and Ga.

[68]

HOLLY FAMILY (Aquifoliaceae)

The word "holly" brings to our minds visions of red berries among thick, evergreen, spiny leaves, associated with the white berries of mistletoe, wreathes of running-cedar, and festivities of the happy Christmas season. The holly referred to above is our American holly (*Ilex opaca*). Known mostly as a small tree it may grow up to 30′ tall and more than a foot in diameter. The small, greenish-white flowers are unisexual, separate on different trees, so that in order to get berries on the female tree it is necessary that a staminate tree be in the vicinity. Fruits are usually red and globose, but in one form they are ovoid and orange or yellow. Fla. to Tex., n. to Mass.

Ilex opaca -white-fruit red

DAHOON, CASSINE (*Ilex Cassine*) A shrub or small tree with relatively large, elliptic, leathery leaves, growing in hammocks, swamps, and bays from Fla. to La. and s.e. Va. This is very ornamental but requires moist, acid soil. It is sometimes also called "yaupon."

Ilex Cassine -white-fruit red

INK- or GALL-BERRY (*Ilex glabra*) The ink-berries are species of holly which have black fruits, leathery, evergreen leaves with few or no spines. They are typically shrubs although some of them will attain a height of 25′. *I. glabra* is the most widespread, occurring in low pinelands, prairies, and borders of swamps of the Coastal Plain from Fla. to La. and n. to N.S. An interesting form of this species occurs in peninsular Fla. with red pointed berries. Besides the above types of hollies, there are a few deciduous species in the South also.

YAUPON (*Ilex vomitoria*) A shrub or small tree up to 26′ tall, with relatively small, leathery, elliptic leaf-blades, crenate-serrate on margin, growing in sandy soil mainly in coastal areas. The leaves and twigs contain the stimulant caffeine, and the Indians made a decoction from them which was used as an emetic before a feast. Hence the specific name *vomitoria*. A tea made from the leaves has been used by coastal people since colonial times. It is one of the most ornamental of the evergreen, red-berried hollies. Under cultivation, it thrives a considerable distance from the coast. Fla. to Tex., Ark., and Va.

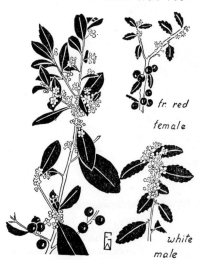

fr. red

female

white

male

Ilex glabra- fr. black -*I vomitoria*

orange-red fr.
red inside capsule

Euonymus americanus-greenish

seeds

Dodonaea viscosa

white

Cardiospermum Halicacabum

STAFF-TREE FAMILY (Celastraceae)

This is a family of several interesting and ornamental shrubs, trees, and vines with simple, alternate, opposite or whorled leaves. Flowers are relatively small and inconspicuous but the fruits and seeds are often showy. Examples are bittersweet, euonymus, and the less known Christmasberry (*Rhacoma*) of s. Fla.

STRAWBERRY-BUSH (*Euonymus americanus*) This erect or straggling shrub presents one of the showiest fruits of autumn in many parts of the South. It is a somewhat fleshy, warty capsule, dark-red and drooping, which splits open into 4 persistent valves, exposing the attached, reddish seeds. It has a weak, 4-angled, green stem and opposite, oval to lanceolate, crenate-serrate leaves which are mostly deciduous. Highly ornamental in gardens or when cut. Various situations, Fla. to Tex. and northward.

VARNISH-LEAF FAMILY (Dodonaeaceae)

VARNISH-LEAF (*Dodonaea viscosa*) A shrub up to 30′ tall, with wedge-shaped, alternate evergreen, sticky leaves. Flowers, without petals in racemes. The winged fruits about 1″ long in shades of red and bronze, borne in showy bunches are both interesting and highly ornamental. Woods and hammocks of peninsular Fla. and tropical Americas.

SOAPBERRY FAMILY (Sapindaceae)

HEART-PEA, HEART-SEED, BALLOON-VINE (*Cardiospermum Halicacabum*) Heart-seed is a translation of the generic name *Cardiospermum* given to these herbaceous or woody vines in allusion to their black seeds with a white spot fancied to be heart-shaped. Flowers are white with unequal sepals and petals and 8 unequal stamens. The fruit is a bladdery capsule. Besides this introduced and cultivated species which tends to escape there are 2 native species in s. Fla. Aside from its ornamental value, the roots of the heart pea are used medicinally.

[70]

BUCKTHORN FAMILY (Rhamnaceae)

A family of shrubs, trees, and woody vines related to the grapes, with 4-5 stamens opposite the petals which are sometimes conspicuously hooded or concave.

SUPPLE-JACK, RATTAN-VINE (*Berchemia scandens*)
The common names of this twining vine emphasize its slender, cord-like stems and its vigorous climbing habit. The neat, elliptic to ovate-lanceolate leaf blades on short petioles have conspicuous, regularly parallel veins. The 5-petaled greenish-yellow flowers are borne in terminal racemes and the berry-like drupes are ellipsoidal and blue. Pinelands and swamps of the Coastal Plain and adjacent areas, Fla. to Tex., Mo. and Va.

greenish-yellow   fruit dark-purple

Berchemia scandens

RED-ROOTS (*Ceanothus*) Mostly low, deciduous shrubs with small, white flowers in umbel-like clusters, the petals conspicuously hooded and stalked (clawed), the calyx and pedicels colored like the corolla. The most common and widespread red-root is New Jersey tea (*C. americanus*). Its leaves were used in colonial times for making tea and the bark and roots are used medicinally. Dry, open woods, various districts, Fla. to Tex. and northward. Southern red-root (*C. intermedius*) is similar to New Jersey tea but with smaller leaves and flowers. Pinelands, Fla. to La., n. to Ga. and Tenn. Small-leaved red-root (*C. microphyllus*) differs from the above 2 in having smaller, oval to orbicular, entire leaves and sessile flower clusters. Dry pinelands and sandhills, Coastal Plain, Fla. to Ala. and Ga.

white

Ceanothus intermedius   C. microphyllus

GRAPE FAMILY (Vitaceae)

A family of woody vines with branched tendrils of modified leaves. Flowers small and variable as to sex. Fruits are usually 4-seed berries. Here belong our many wild grapes for which N.A. is noted.

VIRGINIA-CREEPER (*Parthenocissus quinquefolia*)
As its specific name indicates, this vine has 5-foliate leaves which are palmately compound, but may vary in number from 3-7. The tendril branches terminate in an adhesive disk which enables it to climb flat surfaces. The inflorescence consists of from 25-200 or more flowers in panicled groups of cymes. Berries are deep-blue, pulpy and not edible. Various situations in various provinces, Fla. to Tex. and northward.

[71]

Parthenocissus quinquefolia   berry deep-blue

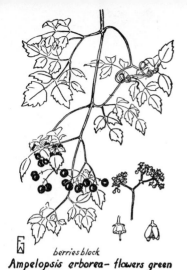

PEPPER-VINE (*Ampelopsis arborea*) This differs from Virginia-creeper in having twice-pinnately compound leaves without adhesive disks. It may be recognized at some distance by its reddish young foliage. Flowers are small and green. Berries, black and inedible. Hammocks, low thickets, and swamps, Coastal Plain, but coastal northward and rarely in other districts. Fla. to Tex., Mo. and Va.

berries black
Ampelopsis arborea– flowers green

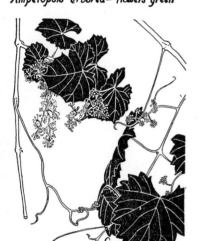

GRAPES (*Vitis*) The best known of the southern wild grapes is muscadine (*V. rotundifolia*), the ancestor of the widely cultivated scuppernong and its varieties. Since this grape is unisexual it is necessary to have a male vine in the vicinity for pollination. Closely related to this is the bullace grape (*V. Munsoniana*) of Fla. and Ga. It differs from muscadine in its smaller, thin-skinned berries with a tender, acid pulp. Mustang-grape (*V. candicans*) of Ark., Okla., and Tex. is represented in Fla. and the W.I. by the variety *coriacea* known as leather-leaf or caloosa-grape. It has toothed, angularly lobed leaves, densely white-hairy beneath. Several other species of grapes occur in the South, the most note-worthy of which is fox-grape (*V. Labrusca*), the ancestor of several well-known vineyard grapes, including Concord, Worden, Hartford, Niagara, and Catawba.

Vitis candicans–V. Munsoniana-white

MALLOW FAMILY (Malvaceae)

SALT-MARSH or COASTAL MALLOW (*Kosteletzkya virginica*) Anyone who has visited the coast dur-ing the summer months has undoubtedly seen the pink, mallow-like flowers in brackish marshes between the dunes and along borders in inlets among salt-marsh rushes and saw-grasses. The foliage is grayish-pubescent with stellate hairs which in variety *altheaefolia* is distinctly velvety to touch. It blooms all summer. The species ranges from Fla. to N.Y. but the velvety variety does not extend n. of N.C.

[72]

Kosteletzkya virginica-pink

ST. JOHN'S-WORT FAMILY (Hypericaceae)

Plants of this family are easily recognized by their opposite, simple leaves which are punctate with transluscent dots and by their mostly yellow flowers with numerous, conspicuous stamens.

ST. PETER'S-WORT, ST. ANDREW'S-CROSS (*Ascyrum*) Low shrubs, distinguished by their 4-petaled, cross-like corollas and 2 large and 2 small sepals. Seven species are listed in Small's *Manual of the Southeastern Flora.* The most frequent and widespread is *A. hypericoides* with 2 styles and ovate larger sepals. Among the more distinctly southern species is *A. tetrapetalum,* recognized by the larger sepals resembling leaves. Low pinelands and swamps, Coastal Plain, Fla. to Ga.

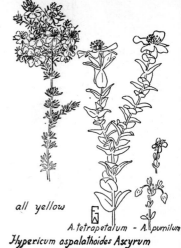

all yellow

A. tetrapetalum - A. pumilum

Hypericum aspalathoides Ascyrum

ST. JOHN'S-WORTS (*Hypericum*) These plants differ from St. Andrew's-crosses in having 5 petals and sepals, the latter not essentially different in size or shape. Among some 30 species in the South, the most ornamental is the golden St. John's-wort (*H. frondosum*). It grows naturally on bluffs and stream banks in the Piedmont and mountain districts from Ga. to Tex., S.C. and Tenn. It is, however, widely cultivated beyond its natural range. A more upright and shrubby species, with numerous small leaves and clusters of smaller leaves in their axils and bright yellow flowers is *H. aspalathoides* which frequents pinelands and prairies from Fla. to La. and N.C.

Hypericum frondosum - yellow

TURNERA FAMILY (Turneraceae)

PIRIQUETAS (*Piriqueta caroliniana*) The Piriquetas are tropical and subtropical herbs with bright to deep-yellow petals and spherical, green capsules of 3 cells in the axils of the upper leaves. Four species occur in Fla., only one (*P. caroliniana*) of which extends as far north as N.C.

ROCK-ROSE FAMILY (Cistaceae)

CAROLINA ROCK-ROSE (*Helianthemum carolinianum*) An interesting characteristic of many rock-roses (also called "frost-weeds") is that, as in many violets, they bear 2 kinds of flowers. The first are showy with petals and the later apetalous. The petaliferous have yellow petals which stay on but a day and many stamens, while the apetalous have few stamens. The showy flowers open only in sunshine. The Carolina rock-rose grows in dry pinelands of the Coastal Plain n. Fla. to e. Tex. and N.C.

yellow

stigma lobed

Helianthemum carolinianum Piriqueta

Stewartia
or
Stuartia Malachodendron-white
red
stamens

TEA or CAMELLIA FAMILY (Theaceae)

This is a remarkable family of trees, shrubs, and woody vines which includes, besides tea, many beautiful ornamentals, especially those belonging to the genus *Camellia*.

STUARTIAS (*Stewartia*) Two species of the genus *Stewartia* are native to N.A. and both occur in the South. Virginia-stuartia, or silky-camellia (*S. Malachodendron*) is a shrub, 3′-15′ tall with alternate, elliptic to ovate leaves, and gorgeous, creamy-white flowers 3″-4″ across, and stamens with reddish-purple filaments and bluish anthers. It is called "silky" from its silky-hairy flowers and under side of leaves. A very attractive shrub which is rarely cultivated. Rich woods, Coastal Plain, and adjacent areas, w. Fla. to La., Tenn., and Va. Mountain-stuartia (*S. ovata*) resembles the above but differs in its separate styles and beaked fruits. Rocky soil, upland districts, Ga. to Ala., and Va.

Franklinia alatamaha - white

FRANKLINIA, LOST-CAMELLIA (*Franklinia alatamaha*) This remarkable small tree was discovered by John and William Bartram on the Altamaha River, s.w. of Ft. Barrington, Ga. in 1765. Since 1790 repeated search has failed to relocate it in its native locality. However, enough seeds and plants had been collected by that time so that it has been preserved through cultivation. It was named in honor of Benjamin Franklin. Resembles loblolly-bay but has thinner, deciduous leaves. Groups of sessile flowers with unequal, white petals appear in July. It is well adapted to a variety of situations and should be cultivated more widely than it is.

Gordonia Lasianthus-white

LOBLOLLY-BAY, TAN-BAY, BLACK-LAUREL (*Gordonia Lasianthus*) A small to a fairly good-sized tree growing in borders of swamps and bays of the Coastal Plain from Lake Okeechobee, Fla. to La. and N.C. It differs from franklinia mainly in its thick, evergreen leaves and pedicelled flowers with united filaments. Its bark is sometimes used for tanning and its heartwood for cabinet-work. Its large, waxy-white flowers commence opening in late spring and continue all summer. It is surprising how far from its natural habitats and range this beautiful, evergreen tree may be grown.

[74]

Violets are among our most popular garden and wild flowers. They belong to the family Violaceae of which there are 15 genera and about 350 species widely distributed over the land surface. They are predominantly herbs but in tropical regions may be shrubs or even trees. In the southern states, there are two genera, *Viola,* or violets, and *Hybanthus,* known as green-violet. The latter is a very different-looking plant from violets, a relatively tall, leafy plant bearing several, small greenish flowers in the upper leaf axils. All of our violets are herbaceous perennials or annuals. Although the name "violet" may suggest that all are of this color, other colors, such as shades of blue, yellow, and white are represented. Another type with 2 of the petals colored differently from the other 3, as in the typical bird-foot violet, occurs in certain species. The cultivated pansy is an exotic, annual violet which through selection and hybridization has given rise to an apparently endless number of forms and varieties of color. All of our native species are perennials, but two introduced and naturalized species occur in the southern flora mostly as harmless weeds. One of these is the field-pansy (*V. Kitaibeliana* var. *Rafinesquii*), locally called "Johnny-jump-up," which sometimes colors the lawns a creamy-white or light-blue. The other is the less common yellow field-pansy of Europe (*V. arvensis*) which has yellow flowers. Violets are commonly classified as "stemmed" and "stemless" meaning that in the former the stems are above ground and in the latter the stems are slightly below the surface as in rootstocks of iris. Most violets have flowers of 2 kinds, the showy, petaliferous flowers which appear in the spring and bear few if any seeds, and the apetalous, inconspicuous ones which appear later in the season and are more prolific in seed production. Many of the so-called blue-violets are notoriously difficult to classify. Few systematic botanists are authorities on violets and many prefer to avoid them. This is because of their high degree of variability; often no 2 plants of the same species look exactly alike. The greatest authority on native violets of North America was the late Ezra Brainerd of the University of Vermont who spent most of his professional life studying violets. He found that the reason for their variability can be attributed largely to the frequency of hybridization in nature. In general, a hybrid is intermediate between the parents with a marked degree of vigor but weakened in seed production. A few years ago, his daughter, Viola Brainerd Baird, published an attractive revision of her father's work, *Wild Violets of North America* (1942) .

purple × lav.   *lavender*

1   2

1Viola pedata 2 var concolor

1

2

1Viola septemloba 2V.triloba-lavender

Viola papilionacea var Priceana-greyish

VIOLET FAMILY (Violaceae)

BIRD-FOOT VIOLET (*Viola pedata*) One of the most beautiful and best known of our southern violets, with leaves divided into narrow, finger-like lobes. Petals are large, flat, all pale to deep lilac-purple or the 2 upper petals differently colored from the others, usually a dark violet. The orange anthers are conspicuous in the center. It is one of the so-called "stemless" violets, meaning that the plant has no stem above ground but a short, thick rootstock, as in iris, on which the flower stalk is borne. It grows in dry, open, sandy or clayey, acid soil in various districts and is widely distributed in e. U.S. from Ga. and Ala. northward.

THREE-LOBED VIOLET (*Viola triloba*) A large-flowered, blue violet with hairy leaves of which the early ones are undivided and the later divided into 3 lobes which may be further divided into shallow lobes. Dry to moist sunny situations in the Piedmont and adjacent areas, Fla. to La. n. to Mo., Ind., and Pa.

SEVEN-LOBED VIOLET (*Viola septemloba*) This differs from the three-lobed in having smooth stems and leaves. The leaves are cordate at base and many are undivided to variously deeply cut. Petals are blue-violet with a white center, and the flower stalks are usually longer than the leaves. It is typically a Coastal Plain species, occurring in sandy-woods, clearings, and edges of swamps. Fla to La. and s.e. Va.

WOOD- or MEADOW-VIOLET (*Viola papilionacea*) Another well-known and widely distributed violet, this form often grows spontaneously at base of shrubbery and other shaded situations around residences. In some localities it is so abundant on wide river bottoms as to form solid patches of deep violet in the spring. A form of this species (var. *Priceana*) sometimes considered a distinct species, with white or grayish petals and prominent blue veins, is known by the name of "Confederate-violet." It is not known in the wild state but is prolific in cultivated or otherwised disturbed ground. Widely cultivated in the southern states but tends to be weedy.

[76]

YELLOW-VIOLETS (*Viola*) Most of the yellow-violets are in the mountain districts. The most common of these is "round-leaved violet" (*V. rotundifolia*) which is as ornamental for its attractive, large round leaves as for its small, yellow flowers. This is also a stemless violet with the leaves close to the ground. Ga. and Tenn. northward. Hastate-leaved violet (*Viola hastata*) is another yellow violet, fairly frequent in the upper Piedmont and the mountains. It is of the "stemmed" form with attractive leaves often variegated with paler green along the principal veins. It grows in rich, hardwood slopes from Fla. and La. to Pa. and Ohio.

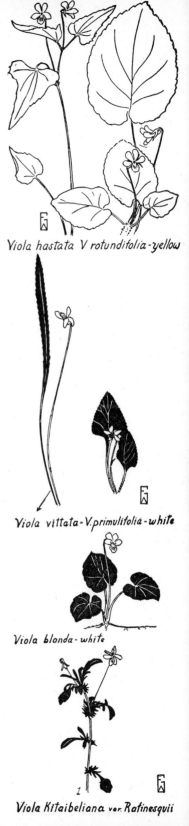

Viola hastata V rotundifolia - yellow

WHITE-VIOLETS (*Viola*) These are most frequently seen in the Coastal Plain, growing in boggy situations where the soil is moist and acid. Primrose-leaved violet (*V. primulifolia*) is generally the most frequently seen white violet of the Coastal Plain, with broadly triangular leaves which may be smooth or hairy. Extends into Piedmont, Fla. to Tex., n. to Okla. and N.J. Lance-leaved violet (*Viola lanceolata*) is very similar to the primrose-leaved violet but has narrower leaves and is always smooth. The longest- and narrowest-leaved violet of the Coastal Plain is *V. vittata* which is so closely related to the lance-leaved that it is questionable if it should be considered a distinct species. Fla. to Tex. n. to e. Can. The common white violet of the mountains is pale-violet (*V. pallens*) with roundish leaves.

Viola vittata - V. primulifolia - white

FIELD-PANSY (*V. Kitaibeliana* var. *Rafinesquii*) A small annual, related to pansies, with bluish-white to creamy petals. A delightful little plant blooming in early spring, frequently in disturbed grounds, such as road-shoulders, borders on fields, and lawns. It may become so abundant on lawns as to form extensive bluish-white patches. It is occasionally associated with the adventive European field-pansy (*V. arvensis*). Named for Constantine Rafinesque (1783-1840), an eccentric French naturalist and botanist explorer of e. U.S. Naturalized from Eurasia and widely distributed in e. U.S.

Viola blanda - white

[77]

Viola Kitaibeliana var. Rafinesquii

Passiflora incarnata - lavender

Opuntia austrina - yellow

female
male
Lindera Benzoin - yellow

PASSION-FLOWER FAMILY (Passifloraceae)

MAY-POP, PURPLE PASSION-FLOWER (*Passiflora incarnata*) In the unique form of the passion-flower, the religious members of the early Spanish explorers of the New World imagined they saw the symbols of the sufferings of Christ. The interesting features of the flower are its stalked pistil, the horizontal, pendant anthers and the fringed "crown" above the petals. The fruit is a large, pulpy berry which is edible. In open ground and various provinces in e. U.S. In the South the most widespread species is the yellow passion-flower, a slender vine, with blunt, 3-lobed leaves and small, yellow flowers.

CACTUS FAMILY (Cactaceae)

Most members of the cactus family are recognized by their leafless, jointed, spiny stems and relatively large flowers with many petals. They originated in the Western Hemisphere but at present are represented in different parts of the world.

SOUTHERN PRICKLY-PEAR (*Opuntia austrina*) A jointed cactus with broad, thinnish "joints" firmly attached, deep or bright-green up to 3′ tall. The spines are slender, yellowish to reddish. Corolla, 2″-3″ broad, light to pale-yellow, petals relatively few. Berry is narrowly obovoid, purple, and edible. Pinelands of s. Fla. Some 30 other species of *Opuntia* occur in the southern states, many of which are widely distributed.

LAUREL FAMILY (Lauraceae)

A family of aromatic shrubs including the camphor-tree, avocado, sassafras, and red- or sweet-bay.

SPICE-BUSH, BENJAMIN-BUSH (*Lindera Benzoin*) A shrub or small tree, bearing dense clusters of small yellow, fragrant flowers before the leaves unfold. Male and female flowers are usually borne on different plants. The red, oblong or oval berry-like fruits, about half an inch long, are very showy in autumn. Wet woods and stream banks, widely distributed in e. N.A. The bark and twigs are used medicinally and the fruits as a condiment. The "hairy spice-bush" or "Jove's-fruit" (*L. melissaefolium*) is more showy with larger flowers, blooming as early as February. Swamps and pond margins, Coastal Plain, Fla. to La. and n. to N.C. and Ill.

[78]

MEADOW-BEAUTY FAMILY (Melastomaceae)

TETRAZYGIA (*Tetrazygia bicolor*) Only 2 genera of this tropical S.A. family are represented in our area. One of these is the herbaceous *Rhexia* and the other the woody *Tetrazygia*. The latter is represented by one tropical species which is either a shrub or a small tree depending on the situations where it grows. It has elliptic leaves, silvery beneath. The white flowers are odd with their one-sided, curved stamens; fruit, a purple or black berry. An unusual and ornamental plant which may be cultivated in warm climates.

Tetrazygia bicolor - white

MEADOW-BEAUTIES, DEER-FLOWERS (*Rhexia*) About 13 species of these herbaceous perennials with opposite leaves and interesting showy flowers occur in the southern states. The petals range in color from pale to bright-purple and one is yellow. The stamens are conspicuous with their strongly asymmetric anthers shedding pollen through pores at the top. Unfortunately, the petals fall readily. The fruit is a vase-like capsule. Savanna meadow-beauty (*R. Alifanus*) is the largest and most beautiful of our meadow-beauties, with a relatively tall, sparingly branched stem terminated by several bright-purple flowers. Where abundant, it colors the landscape during the summer months. N. Fla. to La. and N.C.

magenta
Rhexia Nashii   *R. Alifanus*

MANGROVE FAMILY (Rhizophoraceae)

MANGROVE, RED-MANGROVE (*Rhizophora Mangle*) This large, evergreen shrub is one of the wonders of the plant world. It grows on the borders of salt to brackish shores of tropical lagoons and estuaries, appearing as if standing on its numerous, stilted roots to which its generic name alludes, meaning rootbearing. Small, yellow 4-petaled flowers are borne the year around, the fruit of which bears one seed which grows out into a long, club-shaped seedling before it falls. The submerged parts of the roots are usually covered with small seaweeds and lower forms of marine animals. The bark is used for dyeing and tanning, and the close-grained wood is used for cabinetwork. Fla. Keys, W.I. and tropical S.A.

[79]

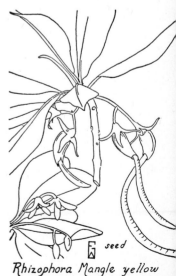

seed
Rhizophora Mangle yellow

ROYAL POINCIANA (*Delonix regia*) Pea Family

The smooth grey trunk of this handsome tree carries a fire of bloom 30'-40' aloft. Even when briefly leafless, the umbrella shape is decorative; but in late May its giant racemes of vermilion blossoms transform the whole tree into a blaze of living fire. Numerous red stamens protrude from the 5 deeply-clawed petals spreading 4" in diameter and revealing thick red sepals with yellow-green backs. One petal is ornately marked, white and yellow blotched with maroon; this darkens the second day, curling and falling to leave a 4-petaled effect. Blossoms rest in a bed of delicate, plume-like leaflets of yellow-green. Native of Madagascar, royal poinciana has become naturalized in the Caribbean area where its aggressive roots constantly break up sidewalks. In June, Miami, Palm Beach, and Ft. Myers are wonderlands of this royal flame flower. Zone 1 and spots of 2. Frost-free and protected citrus areas. Seeds are hard to germinate, so soak overnight in hot water. Transplant young plants in November or December. San Diego, Calif. residents will find the tree grows well there, but it usually fails to bloom. Cut in cool of day, plunge in deep water, submerging leaves for a few hours to prevent wilting. One spray gives a spectacular effect; float individual blossoms.

Royal Poinciana

Golden-Shower

Yellow Tabebuia

Turk's-Cap Mallow

Climbing-Lily

GOLDEN-SHOWER *(Cassia Fistula)*   Pea Family

In early spring, racemes a foot long of claw-like blossoms, like a golden shower, hang gracefully from this small tree. The 5″ long petals open to disclose 5 long stamens and a curious sickle-shaped pistil, which develops into a 1′-2′ long cylindrical pod, well-known for its purgative properties. In its native land, India, the bark of the tree is used medicinally and for tanning. Tender (Zone 1, spots of 2). From seed. Arid regions or well-drained soil by sea shore. Full sun. Keep pruned.

YELLOW TABEBUIA *(Tabebuia argentea)*   Bignonia Family

Terminal panicles of brilliant yellow 2″, tubular flowers of typical bignonia shape literally cover the decorative nude branches of this medium-sized tropical tree from late winter into a long spring season, as the old foliage drops and the new is appearing. The yellowish-green leaves are palmately divided with 5 to 7 pointed leaflets. *T. argentea,* a gnarled, sprawling tree, is used extensively in s. Fla. Zone 1 and most of 2. Propagate by cutting and air layering. Likes rich soil.

TURK'S-CAP MALLOW *(Malvaviscus arboreus* var. *penduliflorus)* Mallow Family

Turk's-cap is an evergreen shrub, growing from coast to coast, which is constantly covered with crimson flowers, hanging fuchsia-like, stamens protruding. The flowers resemble a fez, thus the name "Turk's-cap." *Malvaviscus* means "sticky-mallow," alluding to the young, sticky berry-like fruit. The foliage of yellow-green, heart-shaped, toothed leaves makes it a popular filler shrub. With severe and frequent pruning, sprawling Turk's-cap becomes an effective hedge. Zone 2, part of 3. Grows easily from cuttings. If frozen to ground, recovers quickly even after 20°. Does well in dry or moist soil, sun or shade, resists starving. Flowers remain half-open when cut and last well with no water.

CLIMBING-LILY *(Gloriosa Rothschildiana)*   Lily Family

There is superlative appeal in the dainty but large claw-like blossoms, which, on first opening, are bright yellow, tipped with red, and as they age, turn through orange to garnet. Blooming from spring through fall, when annuals are scarce, *Gloriosa Rothschildiana* is greatly prized for cut flowers and corsages. *G. superba* is a higher climbing, smaller flowered, fall blooming variety. Zones 1, 2, and part of 3. Plant tubers horizontally in full sun; give support; reaches 8′ to 15′. Dig carefully periodically. In north, start in pot, set out after frost. Long lasting in arrangements.

[81]

yellow

Jussiaea scabra   J.peruviana

yellow
Oenothera fruticosa

Oenothera speciosa Oenothera laciniata

EVENING-PRIMROSE FAMILY (Onagraceae)

A family of mostly herbs with alternate or oppo-site leaves and flower parts usually in 4's, the petals often yellow. Fruits, an inferior capsule.

PRIMROSE-WILLOWS (*Jussiaea*) These perennial herbs, or partly woody plants, with axillary yel-low or white flowers were named for Bernard de Jussieu (1699-1776), the gardener of Marie An-toinette and the founder of the first "Natural System of Botany." They inhabit mostly marshy ground. The species with the largest flowers, up to 2½" across, is *J. peruviana*, commonly seen during the summer months in the marshes of s. Fla. but ranges sporadically near the coast as far north as N.C. *J. scabra*, of s. Fla. and the W.I., is distinguished by its broad leaves and bristly, hairy stems. Both are attractive in ponds and fountains.

EVENING-PRIMROSES (*Oenothera*) The evening-primroses are among the showiest of the southern native flowers and deserve to be better known and more commonly cultivated in our gardens. One of the showiest and most frequent and wide-spread is sun-drops (*O. fruticosa*), with large, bright-yellow flowers sometimes 2" across. It grows in various situations from the coast to the mountains from Fla. to Tex. and n. to N.Y. Less showy but equally interesting is the seaside eve ning-primrose (*O. humifusa*) which thrives natu-rally in beach sand and on dunes. Its flowers fade to red. Fla. to La., n. to N.J. Closely related to this is a weedy, winter annual (*O. laciniata*) growing in fields and gardens throughout e. U.S. It is often so abundant as to color the unplowed fields yellow. The most ornamental is the pink evening-primrose (*O. speciosa*), native of La. to Ariz. and n. to Mo. and Kans. It is widely culti-vated and is rapidly becoming naturalized far beyond its natural range, especially in the East and Northeast. Evening-primroses are not re-lated to the primroses of Europe.

[82]

DOGWOOD FAMILY (Cornaceae)

FLOWERING-DOGWOOD (*Cornus florida*) The botanical name *Cornus* comes from a Latin word meaning horn in allusion to its hard wood, and the common name "dogwood" is supposed to have come from the English word "dagwood," a shortening of the expression "dagger-wood," since the wood was used for daggers (skewers) by butchers. Groups of small, pale-yellow flowers are borne at the tips of the branches, covered in bud by 4 large bud scales which expand into 4 large, white or pink petal-like bracts which make the inflorescence of the flowering-dogwood so showy. A form with 6-8 bracts is known as forma *pluribracteata*. The pink-dogwood is botanically known as forma *rubra*. The decorative, elliptic fruits are usually red but may be yellow.

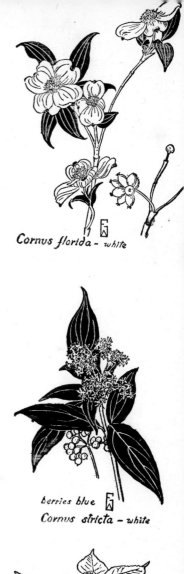

Cornus florida – white

CORNEL (*Cornus stricta*) Most dogwoods do not have the inflorescence of the flowering-dogwood, but small, 4-petaled flowers in flat-topped cymes, resembling those of Viburnums which, however, have 5-petaled flowers. Cornels are mostly shrubs but may attain small tree size. They grow mostly along stream banks, associated with alders and willows. *C. stricta* is a common species in the southern states and by some authors is merged with *C. foemina*. The white flowers with bluish anthers are distinctively ornamental, as are also the bluish drupes. Fla. to Tex., n. to Ind. and Va.

berries blue

Cornus stricta – white

GINSENG FAMILY (Araliaceae)

GINSENG, SANG (*Panax quinquefolius*) Our native ginseng has suffered to a degree approaching complete extinction by gathering the plants ("sanging") and shipping them to China where they are used in superstitious practices. The superstition is based upon the shape of the roots which has a more or less fancied resemblance to man. Because of the lucrative prices for ginseng roots, the natural supply is now practically depleted, and it is, therefore, cultivated locally in mountain districts. It is an attractive plant with a whorl of palmately compound leaves at the summit of a naked stem above which a compact umbel is borne. The fruits are red, flat berries. Cool, rich woods, Fla. to La. and northward.

red berries

male   female

Panax quinquefolius greenish

Eryngium Baldwinii E. aromaticum - blue

pistillate

buds

staminate

Eryngium aquaticum   white

white

Cicuta maculata - Hydrocotyle umbellata

PARSLEY FAMILY (Umbelliferae)

BUTTON-SNAKEROOTS, RATTLESNAKE-MASTERS
(*Eryngium*) Because of their condensed umbels
and simple to slightly compound leaves, Eryn-
giums do not resemble the majority of this family. Corn snakeroot (*Eryngium aquaticum*) has
simple, yucca-like leaves with bristles on the margins and stout, erect stems, 2'-3' tall, branching
near the top and ending in lead-colored heads.
Low woods, swamps and meadows, often in
rather dry soil, Fla. to Tex., n. to Minn. and
Conn. Other species are more slender, with
spreading (*E. aromaticum*) or prostrate (*E. Baldwinii*) stems and small, compound leaves. The
roots of one (*E. synchaetum*) were used in the
ceremonial "black drink" of the Seminoles. The
odd shape and color of their leaves and button-
like heads, as well as their keeping quality, make
them useful in flower arrangements and dried
bouquets.

WATER-HEMLOCK (*Cicuta maculata*) When Socra-
tes ended his life by taking hemlock, he is supposed to have drunk the juice of *Conium*, a plant
which is related to our poisonous hemlock.
Water-hemlock is a tall, robust biennial with
numerous large umbels of white flowers. Its
large, dissected leaves are twice or thrice compound. The purplish-spotted stem gives it the
specific name *maculata*. The tuber-like roots
resembling sweet-potatoes are deadly poisonous.
It is a common plant of marshes and borders of
streams, ponds, and lakes, and is widely distributed in e. N.A., extending as far south as N.C.,
Tenn., and Tex.

MARSH-PENNYWORTS (*Hydrocotyle*) These are
low, attractive plants with extensively creeping
stems and sub-orbicular, shiny leaves and small
umbels of white or creamy flowers. They are
ubiquitous in marshy ground throughout the
Coastal Plain. The most frequent is *H. umbellata*, ranging from Tex. to Mass. *H. bonariensis*
is often seen in large colonies on beach dunes
from Fla. to N.C. A small, delicate, introduced
species called "lawn-pennywort" often invades
moist, shaded spots in lawns and waste places.

[84]

Mock Bishop's-Weed (*Ptilimnium capillaceum*) The attractiveness of this common umbellifer lies in its numerous thread-like leaves and its open, delicate umbels of tiny, white flowers. To really appreciate its ornamental value, it should be potted and brought indoors. In shallow, standing water or in wet soil, frequently seen in borders of marshes, ponds, and roadside ditches. Widely distributed from Fla. to Tex. and northward.

Ptilimnium capillaceum–white

WHITE-ALDER FAMILY (Clethraceae)

Closely related to the heath family but quite distinct as represented in the single genus *Clethra* with some 30 species of which 3 occur in this country.

Sweet Pepperbush, Sweet-Bush (*Clethra alnifolia*) One of our most delightful native shrubs, with its spike-like racemes of white, cylindric flowers which are so "sweet" that they may be detected at considerable distances and, when brought in a room, may be so strong as to be offensive to some. It is the commonest flowering shrub seen in the Coastal Plain during the summer months. Fla. to La. and n. to Me. Another species (*C. tomentosa*) with woolly-hairy leaves, which may even flower in winter, ranges from Fla. to Ala. and N.C. White-alder (*C. acuminata*) is a shrub or small tree of the upper Piedmont and the Appalachian Plateau. Ga. to W.Va. and Va.

Clethra alnifolia - white

WINTERGREEN FAMILY (Pyrolaceae)

Indian-Pipe or **Ghost-Plant** (*Monotropa uniflora*) The nodding, single flower with pinkish stamens on an unbranched, scaly stem strongly suggests a pipe in action. When you are fortunate in finding it in its haunts, enjoy it where it is at home, for soon after being picked it will turn black. It is a remarkable example of a flowering plant completely devoid of chlorophyll and living on humus in association with a mold. False beach-drops and pinesaps are similar plants but lack the pure whiteness of the Indian-pipe.

white
Monotropa uniflora

white
Chimaphila umbellata - C. maculata

SPOTTED-WINTERGREEN, PIPSISSEWA (*Chimaphila maculata*) *Chimaphila* means "winter-loving" which is another way of saying it is "evergreen." It is not true wintergreen which is another plant with small flowers and red berries. Spotted-wintergreen has dark-green, lanceolate leaves variegated with lighter green along the main veins. It is really a tiny shrub which blooms in the late spring and early summer, bearing relatively large flowers (1-5) in a corymb, with reflexed, waxy-white petals suggesting a crown. Dry, acid woods, mostly Piedmont and mountains, Ga. to Ala. and northward. The other species (*C. corymbosa*) with more northern affinities is somewhat rare in the South.

HEATH FAMILY (Ericaceae)

A family of woody plants, noted for their beautiful flowers, many of which are cultivated and through hybridization have given rise to numerous gorgeous varieties.

Befaria racemosa - white tinged pink

TAR-FLOWER, FLY-CATCHER (*Befaria racemosa*) This is one of the most conspicuous flowering shrubs of the Fla. pinelands during the summer months, with its relatively large, white flowers with pinkish tips. The branches are hairy and the buds and calyces are sticky and catch small insects. The 7-petaled corolla is slightly fragrant. A profuse bloomer and when cut as the flower opens, it will keep for several days to a week. An outlying representative of a S.A. genus. Fla. to Ga.

Elliottia racemosa - white

ELLIOTTII (*Elliottia racemosa*) A rare, deciduous shrub named for the noted S.C. botanist, Stephen Elliott (1771-1830). Said by Small to be "one of the rarest of American shrubs." Known only from a few stations in Ga. and s. S.C., represented in each place by only a few individuals or a clump from a common rootstock. As it does not produce seed unless cross-pollinated, it is in danger of becoming extinct. Short panicles bear many white, 5-petaled flowers which bloom in early summer and often last until mid-summer. Only one species is known. Hardy when cultivated as far north as Philadelphia.

[86]

No region of the world, except the Orient, has a more gorgeous assortment of Rhododendrons and Azaleas than the southeastern states. As C. G. Bowers says in his book, *Rhododendrons and Azaleas* (1936), these are "remarkable plants." They occur from the arctic to the tropics in forms from dwarfs to giants. It has been estimated that over 850 species occur in the world, with a countless number of varieties and forms. Also, by artificial hybridization, many horticultural forms have been produced with new combinations of colors and habits and often with an increase of vigor which adapts them better for a wider range of soils and climates. One of their distinctive peculiarities is that they are adapted to a strongly acid soil which is fatal to so many plants, especially those used for horticultural purposes. This acid condition may be an advantage to the plants themselves in reducing competition in nature, but to the horticulturalist it often means the creation of suitable habitats unless the proper situations are naturally available. The southern native Azaleas are usually distinguished from Rhododendrons by their relatively small, deciduous leaves, their distinctly 2-lipped, funnel-form corollas and (usually) 5 stamens, while Rhododendrons have thick, evergreen leaves, more bell-shaped corollas and usually 10 stamens. However, taking the world species into consideration, these distinctions do not hold as too many intergrading forms occur. Consequently, in recent years, there has been a growing tendency to merge the two genera into one; and, as *Rhododendron* has priority, this is the name selected. Despite this, most horticulturalists—and even botanists— will no doubt continue to distinguish Rhododendrons from Azaleas for many years to come. Of our native species, 13 have been classified as Azaleas and 5 as Rhododendrons, while a third, the noted pink-shell azalea (*Biltia Vaseyi*), restricted to a small area in the mountains of N.C., stands out by itself somewhat intermediate between the other two groups. This has probably been used horticulturally more than any of the others in this country. Of the Rhododendrons, *R. catawbiense* is the most ornamental, with its gorgeous clusters of rose- or lilac-purple flowers. Of our Azaleas, the flame- or yellow-azalea (*R. calendulaceum*) of the s. Appalachian Mountains is perhaps the most famous. However, this is equalled in beauty by others, such as the red-azalea (*R. speciosum*) of the Coastal Plain of Ga. and S.C., as well as the Florida flame-azalea (*R. austrinum*) of n.w. Fla.

SAND-MYRTLE (*Leiophyllum buxifolium*) A low to medium tall shrub, with small, elliptic, leathery leaves and numerous white to pink flowers in terminal umbels, with conspicuously red anthers. Where growing abundantly, it presents a showy display in spring. Open, sandy soil, Coastal Plain, N.C. to N.J. A taller form (*L. buxifolium,* var. *Hugeri*) occurs on rocky summits in the upper Piedmont and mountains. Another low, diffuse shrub, locally called "mountain heather," with a profusion of white flowers, is *L. Lyoni* which occurs only on certain rocky summits in the s. Appalachian Mountains.

Leiophyllum buxifolium - white
pink or

SWAMP-AZALEA, SWAMP-HONEYSUCKLE (*Rhododendron viscosum*) This and *R. atlanticum* are 2 of the most common and widespread Azaleas of the Coastal Plain. In spring, there is hardly an open area which does not show the white to pinkish flowers of these two species. Although both are highly variable, they may be distinguished by certain distinct characters. *R. viscosum* does not have underground stems and usually flowers after the leaves appear. Also the corollas in *R. atlanticum* have distinct rows of glands on the outside of the corolla, while in *R. viscosum* they are scattered. Both are strongly fragrant, and when cut they keep well. The latter is not confined to the Coastal Plain as it occurs in "heath balds" at high altitudes in the mountains.

Rhododendron (Azalea) atlanticum-white

FLORIDA FLAME-AZALEA (*Rhododendron austrinum*)This gorgeous species has a limited distribution, being confined to n.w. Fla. It resembles the more northern flame-azalea (*R. calendulaceum*) in flowering as the leaves appear and in the color of its corolla which ranges from light-yellow to deep orange. However, the mountain flame-azalea has a wider range of colors especially toward the reds. The Fla. species differs also from the flame-azalea in having glandless hairs on the corolla. It is slightly fragrant.

[88]

Rhododendron austrinum yellow & orange

PURPLE RHODODENDRON, MOUNTAIN ROSE-BAY
(*Rhododendron catawbiense*) (color page facing
32) This is our most beautiful rhododendron
with its gorgeous rose- or lilac-purple flowers in
large clusters. It has a curious distribution, being found on certain north-facing stream bluffs
in the lower Piedmont but absent at higher elevations except at high altitudes among the
spruce and balsam. It ranges from Ga. and Ala.
to N.C. and W.Va. A few distinct forms have
been recognized. A compact form growing in
the open, especially abundant on Roan Mountain, N.C. has been called var. *compactum,* and
the one in the lower Piedmont, forma *insulare.*
Soon after being discovered, this species was introduced into England where it has contributed
to the origin of many new varieties through hybridization with other species.

Rhododendron catawbiense-light purple

PUNCTATE RHODODENDRONS These are distinguished from the above as well as from the white
or pink great-laurel (*R. maximum*) by their
smaller, glandular or scaly-dotted leaves, especially on the under side, and the smaller rose-
colored flowers. *R. minus* has larger corollas (an
inch or more long) than *R. carolinianum* and
both in turn differ from *R. Chapmanii* in flowering after the new leaves have appeared. The
former occur together in the Blue Ridge but *R.
minus* extends in some areas down to the upper
Coastal Plain. *R. Chapmanii* is restricted to w.
Fla. Albino forms of these, as of other species,
occur.

Rhododendron minus — rose-colored

PINK-SHELL AZALEA (*Biltia Vaseyi*) (color page
facing 32) This exquisite flowering shrub or
small tree combines to some extent the characteristics of both Rhododendrons and Azaleas.
The rose-colored corollas, with yellow or reddish
orange spots, resemble in shape that of a rhododendron, but the deciduous habit is that of the
southern native azalea. The beauty and distinc-
tiveness of this native shrub give it a high ornamental value, and it is, therefore, perhaps more
cultivated than any of the Azaleas and Rhododendrons except *R. catawbiense.* It is also distinctive in its distribution, being restricted to
certain mountain summits of N.C.

[89]

Biltia Vaseyi pink, spots orange

Kalmia latifolia white tinged pink

MOUNTAIN-LAUREL, IVY, CALICO-BUSH (*Kalmia latifolia*) The genus *Kalmia* is composed of 6 species of N.A. evergreen shrubs dedicated to Pehr Kalm (1716-1779), a student of the famous Swedish botanist Linnaeus (1707-1778), who traveled and collected extensively in n. N.A. The corollas are umbrella-shaped with pockets in which the anthers are inserted. Mountain-laurel is the largest species with large, terminal flower clusters, the corollas pink in bud fading white. Extremely floriferous. Dry, rocky woods and stream banks in acid soil, abundant in the mountains but extending along the river bluffs into the lower Coastal Plain. W. Fla. to La. and n. to e. Can.

Kalmia hirsuta - pink

SHEEP-LAURELS, LAMB-KILLS, WICKYS (*Kalmia*) The smaller species of *Kalmia* have smaller leaves and flowers than mountain-laurel and the flowers borne in lateral instead of terminal clusters are in general more highly colored, usually purple to crimson. *K. angustifolia* grows on banks, hillsides, and swamps in various districts from Ga. to Lab. A variety of this (var. *caroliniana*) with grayish, hairy undersides of leaves is more frequent in the Coastal Plain. *K. cuneata*, called white-wicky, has large flowers and wedge-shaped leaves and is confined to acid swamps of e. N.C. and S.C. *K. hirsuta* has hairy calyces and pink or purplish corollas. Coastal Plain, Fla. to Miss. and Va.

Oxydendrum arboreum-white

SOUR-WOOD, SORREL-TREE (*Oxydendrum arboreum*) Most of us are no doubt familiar with sour-wood with its red twigs from which boys make bark whistles, its leaves which are indeed sour to the taste, and its brilliant-red foliage in early autumn. But not everyone has seen or fully appreciates its gracefully upward-arching racemes of small vase-shaped, white flowers which furnish the famous sour-wood honey during the summer months. The close-grained hardwood is used for tool handles. Used horticulturally and should be preserved in wood lots. Widely distributed from the mountains to the Coastal Plain, n. Fla. to La., and northward.

[90]

STAGGER-BUSH (*Lyonia mariana*) At a distance this sometimes resembles honey-cup, but on closer inspection it is seen that the two are quite different as the stagger-bush has cylindraceous corollas, instead of campanulate, and often with a pinkish tint. It is also more widely distributed and inhabits drier soil, growing in acid, sandy or clayey soil in flatwoods and on savannas. Coastal Plain, extending into lower Piedmont, Fla. to Ark., and northward.

FETTER-BUSH (*Lyonia lucida*) This is a beautiful evergreen shrub with thick, shiny leaves, smooth on margin and with a continuous submarginal vein all around the edge. In early spring, the upper parts of the branches are decorated with bunches of usually pinkish, cylindrical flowers which fade white. It is one of the important components of the evergreen shrubby plant communities, the bays or pocosins, of the Coastal Plain. Fla. to La. and n. to Va.

Lyonia lucida-pink-L.mariana white

HONEY-CUP (*Zenobia pulverulenta*) A gorgeous, floriferous, deciduous-leaved shrub with cup- or bell-shaped white flowers some colonies of which are so fragrant that they may be detected by scent at considerable distances. Some plants have whitish underside of leaves while in others they are green. This has been used by some authors as a basis for separating them into 2 species. Damp, sandy, or peaty pinelands and margins of pocosins, n. Fla. to s.e. Va. Should do well when cultivated in damp, acid soil.

Zenobia pulverulenta-white

FETTER-BUSHES (*Leucothoë*) The common name of the evergreen species of this genus alludes to their habit of growing in such thickets that it is difficult to walk through them. Switch-ivy or dog-hobble (*L. editorum*) is one of these which grows in the mountains. The Coastal Plain species, *L. axillaris,* is similar. Both have evergreen, leathery leaves and dense racemes of white urn-shaped flowers. Two deciduous species, *L. recurva* and *L. racemosa,* are very different in appearance from the above, except for the flowers which are borne in terminal, somewhat one-sided racemes.

[91]

Leucothoë racemosa-L.axillaris-white

white to pink-blue berry
Vaccinium
Myrsinites

wax under leaf
Gaylussacia dumosa

BLUEBERRY FAMILY (Vacciniaceae)

A family of berries, including blueberries, huckleberries, bearberries, cranberries, gopherberries, buckberries, and sparkleberies. Flowers are relatively small, mostly urn- or bell-shaped, white or pink. Shrubs or small trees.

DWARF-HUCKLEBERRY (*Gaylussacia dumosa*) Our huckleberries may be distinguished from blueberries by their resinous-dotted leaves which glisten in the sunshine. Also the fruits are 10-seeded while in blueberries they are many-seeded. Dwarf-huckleberry is a common low shrub of dry sandy soil of the Coastal Plain areas, with underground stems, hairy twigs, and slightly glandular-hairy, black berries which are not palatable. Fla. to Miss., n. to N.J.

SPARKLEBERRY, TREE-HUCKLEBERRY (*Batodendron arboreum*) A much branched shrub, or more commonly a small tree, with relatively small, thickish, shiny obovate leaves which tend to persist during the winter months. Flowers are small and bell-shaped, and the berries are small, black, and many-seeded. A form with a whitish bloom is called var. *glaucescens*. In various districts and soil types, Fla. to Tex. and n. to Mo. and s. Va.

red
wine

white
stamens

Polycodium sp.-Batodendron arboreum

GALAX FAMILY (Galacaceae)

GALAX (*Galax aphylla*) Although the leaves of galax have been gathered in large numbers for several generations for the florist trade, it is remarkable how it persists in its frequency and abundance throughout its natural range. And where it is cultivated beyond its range, it tends to escape and become naturalized. The attractiveness of galax is in its relatively large, round heart-shaped, thick, evergreen leaves which keep almost indefinitely after being picked. But many who may be familiar with its leaves have never seen its long, slender, naked flower stalk, terminated by a slender spike of small, white flowers which bloom in late spring. Most common in the mountains but extending down to the Coastal Plain in some sections. Ga. to Ala., n. to W.Va. and Va.

Galax aphylla - white

[92]

SHORTIA, OCONEE-BELLS (*Shortia galacifolia*)
Only 3 species of this interesting genus exist in
the world, 2 in the Orient and one in the gorges
of the s. Appalachian Mountains on both sides of
the border between N.C. and S.C. It has a short
stem with a rosette of leaves with orbicular,
shiny, evergreen blades, crenate-serrate on the
margin. It spreads by short runners and grows
in luxuriant, dense colonies. In early spring it
bears a few white flowers on a slender stalk, with
irregularly toothed petals. It has been cultivated
successfully when transplanted to the right kind
of soil and suitable situations. Named for Dr.
Charles Wilkins Short, a doctor-botanist of Ken-
tucky.

Shortia galacifolia - white

PRIMROSE FAMILY (Primulaceae)

WHORLED-LOOSESTRIFE, CROSS-WORT (*Lysimachia
quadrifolia*) The loosestrifes received their name
from Lysimachus, the King of Thrace who saved
his life from a mad bull by waving a loosestrife
plant before him. They are herbaceous plants
with punctate leaves, as in the St. John's-worts,
which are opposite or apparently whorled. As
the specific name indicates, this species has 4
(sometimes more or less) in each whorl. Yellow
flowers are borne on slender stalks in the axils of
the upper leaves. In moist woods and thickets,
widely distributed, Ga. to Ala. and northward.

yellow

Lysimachia quadrifolia L. terrestris

SHOOTING-STAR, AMERICAN-COWSLIP (*Dodecatheon*) The shooting-stars are so called because of
their nodding flowers with a pointed cone of
joined stamens and the strongly recurved corollas
with long, narrow lobes. They are mostly western but 4 species are found in the East and 3 of
these in South. *D. Meadia* has typically pink
flowers and is northern in distribution but extends down the mountains to Ga. *D. Hugeri*
has white flowers and ranges from Ga. to Ala.,
n. to Ohio. Both are often cultivated in gardens.
The flowers are borne in an umbel at the summit of a naked stalk with a rosette of leaves at
base. The number of flowers is highly variable.

[93]

Dodecatheon Hugeri - white

berry _redish-b_

Jacquinia keyensis -white to yellowish

JOE-WOOD FAMILY (Theophrastaceae)

JOE-WOOD, CUDJOE-WOOD (*Jacquinia keyensis*) A shrub or small tree, the only representative in the South of this family named for the Greek teacher, Theophrastus, known as the "Father of Botany." The opposite, evergreen leaves are thick and very brittle. Small, straw-colored flowers are very fragrant, with 5 petaloid, sterile stamens (staminoides) alternating with the functional ones. Blooming all year. Fruit, an erect, pointed berry which is said to be poisonous like its relatives in S.A. which are used to stupefy fish. Hammocks along the coast, s. Fla. and the Keys.

berry black

Icacorea paniculata - purplish to white

MYRSINE FAMILY (Ardisiaceae)

MARLBERRY, SHELL-MOUND BERRY (*Icacorea paniculata*) A large shrub or small tree, with dense, evergreen foliage and dainty panicles of white to pink bell-shaped flowers, usually blooming in the fall. In spring it is covered with purplish-black berries about the size of a pea, which make it very attractive for birds. Young plants are easily transplanted or it may be grown from seed. Frequent on shell-mounds and in hammocks and pinelands, s. Fla. and the Keys. Good for ornamental planting.

berry black

Bumelia lacuum whitish

SAPODILLA FAMILY (Sapotaceae)

SANDHILL-BUCKTHORN (*Bumelia lacuum*) The buckthorns are mostly evergreen shrubs or trees, usually with thorny branches. Nine species have been reported in the South. The sandhill-buckthorn has leaves, lustrous-white beneath, often becoming brown. Flowers are relatively small with 5 conspicuous staminoids. Berries are on short pedicels, oval and about 1/3" long. Discovered and named by the late John Kunkel Small. Sandhill scrub of c. Fla.

STYRAX FAMILY (Styracaceae)

SILVER-BELL TREES, BELL-TREES (*Halesia*) The silver-bell trees are best recognized by their white, bell-shaped flowers which appear in spring, and by the brown, winged fruits in the fall. The most common species is *H. carolina* which is variously called, "Wild-Olive," "Possum-Wood," "Rattle-Box," and "Calico-Wood." It has a 4-winged fruit, and is widely distributed in woods and on stream banks from the mountains to the lower Piedmont. Fla. to Tex., n. to Ill. and Va. *H. diptera* with a 2-winged fruit, called locally "snow-drop tree" or "cow-licks," has been reported from the Coastal Plain and lower Piedmont from n. Fla. to Tex., Ark., and S.C. A small-flowered species (*H. parviflora*), also with a 4-winged fruit, occurs from n. Fla. to Miss., Okla. and Ga.

white / 4winged seed 2winged seed
Halesia carolina – *H. diptera*

STORAXES (*Styrax*) These are mostly shrubs or small trees that differ from the silver-bells in having deeper lobed corollas with reflexed lobes and capsular, non-winged fruits. *S. grandifolia* has, as the name suggests, large leaves which are white hairy beneath. It occurs in woodlands on certain soil types mostly in the Piedmont. Fla. to La., Ark., Tenn. and Va. A more common species is *S. americana,* often called "Mock-Orange," which occurs along stream banks and in swamps of the Coastal Plain. A form of this with hairy leaves, scurfy beneath, is var. *pulverulenta.* Fla. to Ill. and Va.

Styrax americana – *white*

[95]

For year-round bloom no shrub surpasses the common hibiscus. Spectacular new varieties were developed in Hawaii, but the craze has recently shifted to Fla., s. Calif., and the Gulf states where new crosses of every combination have been developed during the past fifteen years. The American Hibiscus Society lists 607 varieties and hopes to standardize names and classify over 1,000 varieties. The large bouquet illustrated opposite, painted in Fla. in 1936, are Anderson's introductions from Hawaii or his seedlings. Some have proved excellent bloomers, but some are no longer popular. *Lower left, clockwise:* 1-Betty Chalk, 2-*(center)* Anderson's seedling, 3-*(above it)* Sunburst #303 (but similar Red Gold or Macaw bloom better), 4-Urania (a poor bloomer), 5-Sea Cliff, 6-Howard Phipps, 7-Florida Sunset (red with yellow edges), 8-Florida Sunset (good blooming yellow), 9-Ruffled Giant, like Mandalay. Hibiscus shows are increasing in popularity. Tender new varieties are being grown around the Gulf where they are banked with 1'-2' of dirt or mulched with pine straw during cold spells. Many are suitable for pot culture if they can be taken in during freezing weather. They bloom best when grafted on common red stock, but many can be air layered, and some may be propagated by cutting. Several kinds can be grafted on one bush. To help hibiscus fanciers some of the best varieties recommended by Jack Holmes are listed here:

WHITE: Lillian Wilder, Purity, Ruth Wilcox

PINK: Bride, American Beauty, Agnes Gault, Hurricania, Minerva, Mrs. Mary Johnson, Doubles-Kona, Mrs. Mary Morgan, Dainty (La-France)

RED: Single Scarlet, Double Burgundy, Lamberti, Small Schiz., Psyche

YELLOW: Mrs. Earl Anthony, Ruffled Giant, Candlelight, Superba, Y. Mandalay, Y. Buttercup, Y. Delight, Y. California, Double-Butter-fly, Golden Glo-Crown of Bohemia

ORANGE: Hendry's Single Orange, McIntyre Double Apricot, Anderson's 315, Double Flamingo

LAVENDER or VIOLET PINK: Betty Chalk, Kama puaa, Double, Delores Variegated, Fla. Sunset, Jigora, Old Gold

Hibiscus: Psyche, Brilliant, Hurricania

Hawaiian Hibiscus: Jupiter Yellow,
Anderson's #80, Helen Walker

Anderson's Seedling Hibiscus and Early Hawaiian Introductions

Shaving-Brush Tree

Temple-Flower

Jerusalem-Thorn and Bougainvillea

Pink-Ball

SHAVING-BRUSH TREE (*Pachira insignis*)   Bombax Family

Fantastic and odd, almost amusing, are the grotesque patterns made by the sparse thick branches of this small exotic tree of the American tropics when it stands nude of its foliage, with limbs tipped with stiff, cylindrical pinkish brown buds 8″-10″ long, each resembling a stick of shaving soap! Appealing as well, when the buds pop suddenly into bloom, its 5 lingular petals curve miraculously into graceful double rose and brown curls, disclosing hundreds of beautiful, long, showy stamens, reminding one of a huge shaving brush. Zone 1. Frost free areas only. Grown from cutting or from the edible pod-like fruit. Impressive for short-lived arrangements but wilts easily.

TEMPLE-FLOWER (*Plumeria rubra*)   Dogbane Family

This tropical American native is bare in winter, but April fills it with fragrant 8″-10″ branching flower clusters. The leaves soon follow, and the waxy tubular flowers which are 3″ across continue into November. They have the sticky milky juice characteristic of the dogbane family. Varieties from white to yellow, pink, rose-purple, and red. Tender (Zone 1, protected parts of 2). From cuttings. Drought resisting. Buds open in house.

JERUSALEM-THORN (*Parkinsonia aculeata*)   Pea Family

A graceful, small tree distinctive for its feathery, drooping, almost pendulous zig-zagging branches and profusion of yellow flowers borne in spring in a mist of fine green foliage. Zones 1 and 2. From seed. Prune when young. Heat and drought resistant. Lasts well when cut; seed pods for dried bouquets.

BOUGAINVILLEA (*Bougainvillea glabra* hybrids)   Four-o'Clock Family

This is perhaps the most popular woody vine for all warm climates. The magenta-purple *B. glabra* is easily pruned into trees and hedges. The spectacular red, *B. spectabilis* var. *crimson lake,* is a more robust grower. Recent hybrids can be had in almost any shade and hue from white through yellows, oranges, purples, pinks, to deepest red. Zones 1 and 2.

PINK-BALL (*Dombeya Wallichii*)   Sterculia Family

Pink-ball is an apt description of the tight clustered blossoms of this large-leafed sprawling shrub. The flowers are half hidden by the huge heart-shaped leaves. Each blossom resembles a small, wild rose, but they hang pendulous on slender stems, and the old flowers brown as they fade. Zones 1 and 2; parts of 3. Withstands 25° in Fla. to coastal Calif.

OLIVE FAMILY (Oleaceae)

A family composed mostly of woody plants with opposite leaves, including, besides olive, such familiar plants as ashes, lilacs, privets, and Jasminums.

Chionanthus virginicus—white

FRINGE-TREE, GRANDSIR-GRAYBEARD (*Chionanthus virginicus*) A small tree with opposite, simple, deciduous leaves. Very floriferous, the whole tree covered with white, long-linear, 4-petaled flowers, the petals slightly united at base, in drooping, festoon-like panicles. Highly ornamental and easily transplanted when young or planted from seed. A striking combination is to plant coral-honeysuckle at its base, as they bloom at the same time. Various situations, Fla. to Tex., n. to Mo. and N.J.

Osmanthus americanus — white

AMERICAN WILD-OLIVE (*Osmanthus americanus*) Usually seen as a small tree but may become 40′ tall, with narrowly elliptic, evergreen leaves, lustrous above. The small flowers are white to greenish-white. The fruit is an ellipsoidal drupe about an inch long and dark-purple when mature. However, some plants have yellowish-green fruits and may represent a distinct species. A species with larger globose fruits (*O. megacarpus*) is restricted to the southern end of the Fla. lake region. Hammocks, swamps, and bluffs, Coastal Plain, Fla. to La. and Cape Henry, Va.

GENTIAN FAMILY (Gentianaceae)

This family no doubt calls to mind the blue gentians of autumn days, but it includes many other lovely flowers of other colors, which bloom at other seasons.

SEASIDE-GENTIAN (*Eustoma exaltatum*) An annual herb of s. Fla., blooming nearly all year, with broadly bell-shaped, rose-purple or lavender corollas with a dark center, 1″-1½″ across. Leaves, opposite, gray-green, elliptic or oblong. Chiefly near the coast, in hammocks, pinelands, and among coastal dunes.

[98]

Eustoma exaltatum—violet & purple

MARSH-PINKS, ROSE-GENTIANS (*Sabatia*) The Sabatias do not superficially resemble gentians although they belong to the same family. Their corollas are more deeply parted and the color varies from white to pink or red, never blue.

SEASIDE-PINK (*Sabatia stellaris*) This is one of the showiest flowers at the seacoast, growing in brackish marshes but more often seen between and back of the dunes. It is a sparingly branching annual with pink, deep-rose or magenta corollas with a yellow star-shaped center. White-flowered forms occur. La. to Fla., n. to Mass. Elliott's-sabatia (*Sabatia Elliottii*) is a unique species in having a white to cream-colored corolla with mostly wedge-shaped lobes much longer than the calyx. It flowers freely during summer and fall. A southern species ranging in the Coastal Plain from Fla. to Ala. and N.C.

LARGE-FLOWERED SABATIA (*Sabatia grandiflora*) In the low pinelands, prairies, and parts of the Everglades of s. Fla. grows the largest-flowered of the 5-petaled Sabatias. The deep-rose or magenta corolla, up to 2″ across and with a yellow "eye," may be seen nearly all year. It has alternate, filiform upper leaves and a relatively short calyx. It could undoubtedly be cultivated beyond its natural range.

ROSE-PINK, BITTER-BLOOM (*Sabatia angularis*) This biennial with its numerous, deep-pink corollas always attracts attention from roadside clearings and forest margins in late summer and autumn. It is especially noticeable for growing in sterile soil where it is not hidden by many other plants. It should be a beautiful addition to our flower gardens. Widely distributed from Fla. northward. Many-petaled marsh-pink (*Sabatia gentianoides*) is one among 4 of the southern Sabatias with 8-10 lobes of the corolla. They are found mainly in the Coastal Plain. In this species the corollas are reddish, lilac, or pink-purple. Others are deep-rose to white. Fla. to Tex. and N.C.

white male & female stages
Sabatia Elliottii

pink
Sabatia grandiflora

pink
Sabatia angularis & gentianoides

Gentiana Catesbaei- *blue or rose-purple*

CATESBY'S-GENTIAN (*Gentiana Catesbaei*) Except for the fringed-gentian (*G. crinita*) which is rare in the South, most of the southern species have relatively narrow corollas with short, tardily-opening lobes and flower parts in 4's. Eight species have been reported from the South. Catesby's-gentian with a somewhat bell-shaped corolla is named for its discoverer, Mark Catesby (1679-1794). Its color varies from deep-blue to violet purple. It is quite variable and has sometimes been split into more than one species or into several varieties. It grows in moist, sandy soil around peat bogs, in pinelands and wet woods. Coastal Plain, Fla. to Ala. and n. to Del. A form with roundish, blunt leaves has been segregated as var. *nummulariaefolia*.

greenish-white or

Gentianavillosa - *purplish-green*

STRIPED-GENTIAN, SAMPSON'S-SNAKEROOT (*Gentiana villosa*) One of the most widely distributed gentians which is fairly frequent although never especially abundant. Called striped-gentian because of the stripes in the coloration of its greenish-white to purplish-green corolla. It is not hairy as the specific name suggests. Low woods and river bottoms in various provinces. Fla. to La. and n. to Ind. and N.J. The soapwort-gentian (*G. Saponaria*) is so called because its leaves resemble those of bouncing-bet (*Saponaria officinalis*). It differs from striped-gentian in its shorter corolla lobes, hairy-margined sepals and darker blue corolla. In uplands, Ga. northward.

light blue   *Indigo*

Gentiana Saponaria- *G. autumnalis*

PINEBARREN-GENTIAN, SANDHILL-GENTIAN (*Gentiana autumnalis*) This remarkable flower is a surprise to anyone seeing it for the first time. In its open funnel-shaped corolla it resembles more a lily or an amaryllis than a gentian, but the 5 united petals with toothed plaits between the lobes, as well as the rich, indigo-blue color reveal its true relationship. It bears few flowers, often only one, on a slender stem with thickish, long, narrow leaves. Moist, sandy soil, Coastal Plain, S.C. to N.J. A closely related species (*G. tenuifolia*) with white corolla lobes and narrower leaves grows in middle Fla.

[100]

DOGBANE FAMILY (Apocynaceae)

A family consisting of herbs, shrubs, or trees, mostly with milky juice and closely related to milkweeds, but differing from them in the united styles and distinct stamens. Includes such familiar plants as blue-dogbanes, periwinkle, oleander, allamanda, and others.

BLUE-DOGBANE (*Amsonia Tabernaemontana*) The blue-dogbane is an attractive plant with panicles of blue flowers. The generic name was the name of an early physician of Va., Dr. Amson (1760), and the name of this species was dedicated to a German herbalist, Tabernaemontanus, of the 16th century. A perennial herb up to 20″ tall, with alternate, mostly broadly-elliptic-lanceolate leaves. Blooms in spring. Rich woods and floodplains, Fla. to Tex. and Va. Naturalized northward. Should be cultivated more than it is. *A. ciliata* has numerous filiform leaves. Coastal Plain, Fla. to Tex., Ark., and N. C.

WILD-ALLAMANDA (*Urechites lutea*) A gorgeous, high-twining vine, with opposite, flat, bright-green leaves and yellow flowers in axillary clusters. Corollas expanded from a bell-shaped tube about 2″ across. Plants highly variable. Hammocks of s. Fla. and the Keys; also in the W.I.

ROSE-PERIWINKLE, OLD-MAID (*Vinca rosea*) This erect herb which was introduced from Madagascar has become naturalized in hammocks, pinelands, waste places, and roadsides in peninsular Fla. and the Keys. The corollas have a long tube and broad, flat lobes, white, pink, or rose-purple with a reddish eye. Sometimes called Madagascar periwinkle. Extremely abundant in certain sections. Often cultivated as an annual in other sections.

RUBBER-VINE (*Rhabdadenia biflora*) The elongate stems bear opposite leaves, apiculate at tip. Flowers are white with broad calyx lobes and a long corolla tube expanded above, its lobes convolute-imbricate in bud (fan-shaped). Fruit, 2 pods. Coastal hammocks, s. Fla. and the Keys.

[101]

blue

Amsonia ciliata - A. Tabernaemontana

Urechites lutea - yellow

white or rose / white

Vinca rosea - Rhabdadenia biflora

DEVIL'S-POTATO, RUBBER-VINE (*Echites umbellata*)
A vigorous vine, its stems and branches often
intricately intertwined, with opposite, broadly
ovate or oval leaves. The white to greenish-
white flowers have a long tube with an expansion
below the spreading lobes with toothed margins.
The short calyx has glands inside. Hammocks
and pinelands of s. Fla. and the Keys extending
to the W.I.

Echites umbellata   white

MILKWEED FAMILY (Asclepiadaceae)

A family composed of perennial herbaceous or
woody plants usually with a milky juice, opposite or whorled (rarely scattered) leaves and
mostly umbel-like cymes of flowers with joined
stamens and anthers joined to a common stigma
of two pistils of which often only one matures
into a pod (follicle). Highly adapted to insect
pollination so that insects will pull out pollinia
(pollen sacs) with their legs.

MILKWEED-VINE (*Funastrum clausum*) A vine with
ovate-elliptic or elliptic-lanceolate leaves and
long-peduncled, axillary umbels of white fragrant flowers blooming all year. Climbing over
herbs and shrubs, hammocks of coastal and lake
regions of peninsular Fla. and the Keys; also
W.I., Mex. and S.A.

white

Funastrum clausum

MILKWEEDS, SILKWEEDS (*Asclepias*) Erect or
spreading, perennial herbs, a few without milky
juice, classified mainly upon the variation of the
"crown," consisting of hood-like projections outside of the stamens, color of corollas, and characteristics of follicles. More than 20 species have
been reported for the South. Sandhill-milkweed
(*A. humistrata*) is a striking plant with prostrate
stems and erect leaf blades, waxy light-green and
conspicuously white-veined. Corolla lobes gray
or greenish-purple. One of the most distinctive
plants, frequenting dry pinelands, scrub, and
sandhills, Coastal Plain, Fla. to Miss. and N.C.

pink veined leaves

Asclepias humistrata-ashy & purple

[102]

METASTELMA (*Metastelma Blodgettii*) A slender vine which is often seen growing in great profusion over herbs and bushes, with narrow, downwardly directed leaves. Small flowers in sessile, axillary umbels. Corolla white, vase-shaped, the lobes spreading and hairy within; the crown attached to the base of the corolla tube. Hammocks and sand dunes, Fla. to s. Tex.

MORNING-GLORY FAMILY (Convolvulaceae)

The morning-glories are among the most decorative of our wild flowers, but because of the weedy characteristics of some species, such as the common morning-glory (*Ipomoea purpurea*) they are not generally appreciated as much as they should be.

RAILROAD-VINE (*Ipomoea Pes-Caprae*) This tropical and sub-tropical species seems to prefer growing in the most inhospitable situations, creeping along railroad tracks, roadsides, and ocean beaches. The flowers are large and purple; the leaves broad, thickish, shining, notched at apex, cordate at base and often folded as the covers of a book showing the prominent veins on the back. Fla. to Tex. and Ga.; W.I.

ARROW-LEAVED MORNING-GLORY (*Ipomoea sagittata*) The flowers of this species are quite similar to those of the railroad-vine, but the leaves are strikingly different, varying from broad to narrow arrow-shaped. It often inhabits similar situations. It climbs in shrubby thickets on borders of marshes near the coast and is often seen creeping on road shoulders among the dunes. Fla. to Tex. and N.C. White-operculina (*I. dissecta*), as well as the yellow morning-glory (*I. tuberosa*), has sometimes been placed in the genus *Operculina* because of its urn-shaped calyx and twisted stamens. The white-operculina is readily distinguished from all other morning-glories by its conspicuously dissected leaves. Pinelands, Fla. to Tex. and Ga. Seven-leaved morning-glory (*I. heptaphylla*), a beautiful species with usually 5-lobed leaves and pale-violet or lavender flowers has slender peduncles which are often twisted for use in climbing. Naturalized from the tropical regions in s. La.

yellow

white to rose-purple

Ipomoea purpurea   *Metastelma Blodgettii*

Ipomoea Pes-Caprae - mauve

white, throat purple

Ipomoea dissecta — I. sagittata-rose

Calonyction aculeatum - white

MOON-FLOWER, MOON-VINE (*Calonyction aculeatum*) This high-climbing vine grows profusely over burned over areas as a "fire-weed." Its large, white corollas, with a long, slender tube, widely expanded above into a flat top up to 8″ across, open at night. The ovate to orbicular leaves, frequently 3- to 5-lobed at base, are often very large. It is distinguished from its close relative, *C. Tuba,* by its appendaged calyx and its inland distribution. The generic name *Calonyction* means "beautiful night." Both are pleasantly fragrant. Hammocks of Fla. to La.

DODDER FAMILY (Cuscutaceae)

DODDER, LOVE-VINES (*Cuscuta*) Most of us are no doubt familiar with the pale-yellow, stringy growths that sometimes cover vegetation in nature as well as cultivated plants. These are parasitic flowering plants with reduced leaves and groups of small flowers along the stems. If growing in abundance they may be quite destructive to the host plants. Most of these belong to the genus *Cuscuta* related to morning-glories. But a plant of similar habit is found in peninsular Fla. and the Keys which according to its flower characters shows no close relationship to dodders. This is *Cassytha filiformis* which is closely related to plants of the laurel family and has a spicy fragrance. It is quite destructive on rosemary (*Ceratiola ericoides*).

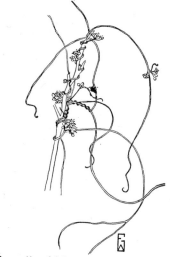

Cassytha filiformis - creamish

WATER-LEAF FAMILY (Hydrophyllaceae)

NAMAS, HYDROLEAS (*Nama, Hydrolea*) Few who are not botanists are acquainted with these attractive perennial herbs which often inhabit marshy areas so heavily populated with sedges and rushes that they are difficult to explore. They have relatively large flowers with delicate petals and attractive globular capsules. *N. corymbosum* has corollas mainly of azure-blue while in *N. ovatum* they are purplish or white. Both may be cultivated successfully. Related to *Nemophila* some species of which are cultivated.

[104]

Nama corymbosum - N. ovatum - blue

PHLOX FAMILY (Polemoniaceae)

STANDING-CYPRESS, SPANISH-LARKSPUR (*Gilia rubra*) This tall biennial with numerous leaves, divided into thread-like divisions, vies with the cardinal-flower in its brilliant, scarlet spikes, and as it grows in open fields and forest margins is more often seen and admired. Its long-tubed corollas with flaring lobes spotted within are an adaptation for pollination by humming birds. During the first year, it produces a rosette of numerous leaves and the following year sends up a tall stem terminating in a long flowering spike. Coastal Plain and adjacent areas, Fla. to Tex. Cultivated and escaping beyond its natural range.

Gilia rubra   red

PHLOXES (*Phlox*) Few native flowers have become more popular as garden flowers than the phloxes. Of these, the annual *Phlox Drummondii* heads the list. Its original home seems to have been in Tex. whence a horticulturalist plant collector, Thomas Drummond, sent seeds to England in 1835. Its outstanding characteristics are its annual habit, its alternate upper leaves, and its sticky hairiness. Under cultivation it has given rise to an innumerable variety of flower colors and patterns which by hybridization have given rise to still further forms. In recent times it has migrated eastward along the coast, growing profusely in waste places of coastal cities as far north as Va., associated with the blanket flower (*Gaillardia pulchella*) with a similar history. Another popular native phlox is moss-pink or subulate-phlox (mainly *P. nivalis*). This is usually transplanted. In early spring, beds of pink and white are often seen bordering lawns and in rock gardens from the humblest cabin to the stateliest mansions. Their distinctive character is the numerous short, evergreen, sharp-pointed leaves. Of the tall species there are several with southern distribution which are also highly ornamental. One of these is smooth-phlox (*P. glaberrima*) with thick, long, narrow leaves growing in various situations but most commonly in the Piedmont. Very similar to this is Carolina-phlox (*P. carolina*) which is usually finely hairy and has broader leaves. Downy-phlox (*P. pilosa*) is closely related to the common blue-phlox from which it differs in its hairiness and absence of runners.

Phlox Drummondii   many hued

[105]

Phlox nivalis – P. pilosa –   pale violet

Physalis maritima Curtis

NIGHT-SHADE FAMILY (Solanaceae)

A paradoxical family of plants some of which are extremely poisonous, such as belladonna, strychnine, and Jimson-weed, while others, such as irish-potato, eggplant and tomato are valuable food plants. Many are highly ornamental.

SEASIDE GROUND-CHERRY (*Physalis maritima*) The interesting and ornamental part of ground-cherries is not so much in its flowers as in its bladdery calyces which enclose a berry. One of the easiest kinds to recognize is the seaside ground-cherry as no other species grows on ocean beaches and sand dunes. Flowers are nodding, greenish-yellow with a dark center, and the fruiting calyx is about 1″ long. The plants are sticky hairy. Fla. to Tex. and Va.

Solanum aculeatissimum

SODA-APPLE, LOVE-APPLE (*Solanum aculeatissimum*) Although this plant belongs to the same genus as irish-potato, it is a very different looking plant. As the specific name suggests, it is extremely sharp-spiny. A perennial, partly woody plant, it bears relatively large, orange to red, papery berries which make up the ornamental dried bouquets often seen in florists' shops in the fall. It may be grown as an annual beyond its natural range if the seeds are germinated indoors in early spring. Sandy fields, thickets, and roadsides, Coastal Plain, Fla. to Tex. n. to s.e. N.C.

BIRD-PEPPER (*Capsicum frutescens*) A woody perennial from 1′-4′ tall, bearing small, white flowers and oblong, red berries which birds are very fond of. It is one of the "hot-peppers" as you will soon realize when you taste the berry. Hammocks of s. Fla. and the Keys.

CHRISTMAS-BERRY (*Lycium carolinianum*) A relative of the plant known as matrimony-vine or box-thorn (*Lycium halimifolium*), this is a spiny shrub with small, fleshy club-shaped leaves and lavender or white flowers with red berries 3″-5″ long. In various coastal situations, Fla. to Tex.. n. to S.C.

lavender or
white   whitish

Lycium carolinianum Capsicum frutescens

[106]

BORAGE or SCORPION-WEED FAMILY
(Boraginaceae)

Certain members of this family are often called "scorpion-weeds" because of their one-sided racemes which are usually spirally coiled, remotely resembling the upwardly curled tail of a scorpion. Mostly known for their ornamental value, such as forget-me-nots, heliotropes, and hound's-tongues.

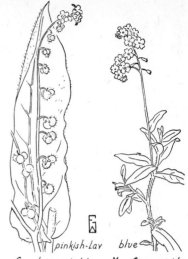

pinkish-lav   blue

Cynoglossum virginianum Myosotis scorpioides

HOUND'S-TONGUE, WILD COMFRY (*Cynoglossum virginianum*) Growing in rich woods in various districts except the Coastal Plain, this plant is recognized by its large tongue-shaped basal leaves, a flowering stem up to 3' tall bearing a group of racemes at the top with light-blue flowers, with a white center, resembling large flowers of forget-me-nots. Fruits are relatively large, spiny, wrinkled nutlets which can be germinated successfully in your garden. This species is as ornamental as some of the cultivated exotic ones. Ga. to La., Kan., and N.J.

VIRGINIA-BLUEBELL, COWSLIP (*Mertensia virginica*) A beautiful, native flower, named for Virginia where it was first found. Perennial herb with a large root and elliptic to oval leaves, often growing in clusters. Flowers drooping, corollas blue, rarely white, with a long tube, cup-shaped above. Scattered in distribution from Ala. northward, but often cultivated within as well as outside of its natural range.

Mertensia virginica -blue, pink tipped

YELLOW PUCCOON, GROMWELL (*Lithospermum canescens*) A very showy, hairy perennial with large, colored roots and orange-yellow corollas with spreading lobes. Distinguished from the closely related *L. carolinensis* by its hirsute instead of hispid (stiff-hairy) stems and beardless corolla tube at base. It grows in rich woods, but more often in gravelly, clayey, poorly drained soil in various districts but perhaps most frequently in the lower Piedmont and Coastal Plain. A distinctive ornamental for the wildflower garden and the rock gardens. Ga. to Tex. and northward.

Lithospermum canescens - yellow

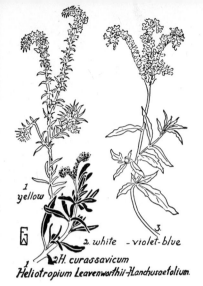

1 yellow

2. white - violet-blue

2H. curassavicum

1 Heliotropium Leavenworthii-Hanchusaefolium.

HELIOTROPES (*Heliotropium*) The name helio-trope, which means sun-turning, came from the ancient belief that when a plant flowers it turns toward the sun.

SEASIDE HELIOTROPE (*H. curassavicum*) This is the most frequently seen heliotrope in the south-ern Coastal Plain. It is a spreading perennial with narrow wedge-shaped leaves and dense-flowering racemes of white flowers with a yellow center, fading blue. A common plant in sandy soil of hammocks, shores, and waste places. Coastal Plain, Fla. to Tex., n. to Del.; tropical America. Leavenworth's-heliotrope (*H. Leaven-worthii*) is a pale-hairy plant with erect stems and yellow flowers, growing in wet places of low hammocks, mostly in the Everglades of s. Fla. and the Keys. Blooms all year.

Tournefortia gnaphalodes- white

TOURNEFORTIA (*Tournefortia gnaphalodes*) A stout, trailing or climbing, woody vine with scat-tered, alternate, usually harshly pubescent leaves and white corollas with flat or crisped, spread-ing lobes. The only southern representative of this genus is named for the noted French bota-nist, Joseph Pitton de Tournefort (1656-1708), often called the "Father of the Genus," the first botanist to attempt to describe systematically genera of plants. Hammocks of s. Fla. and the Keys.

VERVAIN FAMILY (Verbenaceae)

MOSS-VERBENA (*Verbena tenuisecta*) This resem-bles a true verbena in its flower characteristics but because of its flat-topped inflorescence and appendaged stamens, it is sometimes placed in the genus *Glandularia*. A reclining, hairy plant with opposite, dissected leaves with narrow lobes. Flowers are small but showy, with purple, pink, or white corollas. Sandy or clayey soil, roadsides and waste places. Coastal Plain, Fla. to Ga. and La. Adventive from S.A. This is often asso-ciated with another member of the Verbenaceae, *Lippia nodiflora,* commonly called "cape-weed," a trailing perennial with opposite, wedge-shaped leaves and short heads of small purplish to al-most white flowers. Fla. to Tex., n. to Va.

[108]

mauve white to purple.

Verbena Tenuisecta-Lippia nodiflora

SEASIDE-VERBENA (*Verbena maritima*) This is another flat-topped verbena, the only species often frequenting coastal situations of dunes, dry pinelands, and hammocks. Flowers range from rose to bright-purple. Plants spreading to widely creeping. S. peninsula of Fla.

CAROLINA-VERBENA (*Verbena caroliniana*) This is another species which varies somewhat in its flower characters from a true verbena and is therefore sometimes placed in another genus, *Stylodon*. However, the habit of the plant is typical of a verbena. The more or less interrupted spikes bear small, pink, purplish, or white corollas. Sandy soil, Coastal Plain, Fla. to La. and N.C.

white → pink to lavender
purple

Verbena caroliniana V. maritima

LANTANAS, SHRUB-VERBENAS (*Lantana*) Lantanas are closely related to Verbenas but are mostly shrubby, with flat-topped inflorescences. *L. Camara,* which is often cultivated in the Gulf states, has spiny branches. A narrow-leaved form is probably introduced from the W.I., while the broad-leaved form is native. Corollas range in color from yellow to pink, fading scarlet. Sandy soil, Coastal Plain, Fla. to Tex. and Ga. The weeping-lantana or "polecat-geranium" (*L. Sellowiana*) is a shrub with hairy leaves and magenta or lilac corollas. Naturalized from S.A. Coastal Plain, pinelands, roadside, and waste places, Fla.

variegated
lilac

Lantana Camara L. Sellowiana.

BEAUTY-BERRIES, FRENCH-MULBERRY (*Callicarpa americana*) A shrub, 3'-6' tall with opposite, ovate to elliptic leaves and numerous small, bluish or pinkish flowers in congested cymes in the axils of the leaves. The bunches of small, violet or magenta berry-like drupes are the ornamental part of the plant. This native species is more showy than that introduced from Japan which is often cultivated and escapes locally to low woods and stream banks. Mostly in low woods and thickets of the Coastal Plain and adjacent areas, Fla. to Tex., Ark. and s. Va. Attractive for arrangements but soon withers.

Doge purple

Callicarpa americana -pinkish or bluish

Trichostema suffrutescens

This family includes many well-known herbs, such as pennyroyal, hoarhound, sages and catnip, besides many ornamentals.

BLUE-CURLS (*Trichostema suffrutescens*) A somewhat woody biennial or perennial of a number of species called blue-curls because of their bluish corollas and especially because of its downwardly curled, long bluish stamens. Flowers with a very irregular calyx occur in axillary cymes. It grows in dry soils and scrub of Fla. Two annual species of blue-curls of wider distribution are: *T. dichotoma* with broad hairy leaves, and *T. lineare* with smooth, narrow leaves.

bluish to violet   pale blue to white

Scutellaria integrifolia S. multiglandulosa

SKULLCAPS or HELMET-FLOWERS (*Scutellaria*) The skullcaps are distinguished by, and get their name from, an inverted cap-like projection on the upper side of the calyx tube. The entire-leaved skullcap (*S. integrifolia*) is a frequent and widely distributed species, growing in a wide range of situations from the coast to the mountains. It is distinguished by its soft hairiness and smooth-edged upper leaves. The corolla is a light-blue with the upper lip much longer than the lower. Glandular-skullcap (*S. multiglandulosa*) is a more beautiful species with a broad lower lip crisped on the margin and exceeding the upper lip. Corolla is a light-blue to white. As its specific name suggests, the plant is copiously sticky, glandular-hairy. Dry situations, Fla. to Ga.

pink to magenta

Physostegia denticulata

FALSE DRAGON-HEAD (*Physostegia denticulata*) This is an erect, relatively tall perennial with few branches and rather large, light-purplish to roseate corollas which are tubular and somewhat inflated above. The opposite leaves are thickish with distant blunt teeth and the upper are conspicuously reduced. Sometimes called "obedient-plants" because when the flowers are pressed up against the stem they will remain in this position for some time. The more widely distributed *P. virginiana* with thinner leaves sharply toothed on margins is often cultivated and soon spreads beyond its bounds by its aggressive rootstocks. The former occurs in moist soil in various situations in the Coastal Plain, Fla. to Tex., Kans. and Md.

SAGES (*Salvia*) Sages are generally well known, having been cultivated from ancient times as herbs and ornamentals. Even now there is hardly a garden without some variety of sage.

Salvia lyrata - violet-blue

PURPLE- or LYRE-LEAVED SAGE (*Salvia lyrata*) The commonest of our native sages, often appearing in great abundance on borders of lawns and open woods cleared of undergrowth. The principal leaves are mainly basal, usually purplish-tinted. The bluish flowers are borne in distantly spaced whorls on a naked stalk except for one pair of reduced leaves at the base of the inflorescence. Various situations and provinces, Fla. to Tex. and northward.

BLUE-SAGE (*Salvia azurea*) The heavenly-blue flowers associated with the soft, light hairy stems and foliage, have brought this native flower to many of our southern gardens, even beyond the limits of its natural range. It is a perennial with narrow leaves and spikes of remote whorls of azure-blue or white flowers. It grows naturally in dry or sandy soil of various situations in the Coastal Plain, Fla. to Tex. and S.C. Red- or scarlet-sage (*S. coccinea*) is the only red-flowered native sage which is both conspicuous and easily recognized. A soft-hairy perennial with ovate crenate-serrate leaves, it frequents roadsides and waste places in the Coastal Plain, Fla. to Tex., S.C. and Mex.

Salvia azurea - blue   S. coccinea-red

LION'S EAR (*Leonotis nepetaefolia*) A distinct, soft pubescent annual up to 6′ tall introduced from Africa and somewhat naturalized in cultivated ground, roadsides, and waste places from Fla. to La., Tenn., and N.C. The unusual and picturesque feature of this plant is its large, dense, globular cymules of orange-yellow or scarlet flowers around the upper nodes. The upper lip of the corolla is arching and much longer than the lower. The calyces are long, conspicuously curved with lobes ending in stiff bristles. The glomerules of calyces remain on the plant long after the corollas have fallen off and are used in dried bouquets.

orange
Leonotis nepetaefolia "Lion's-Tail"

This group of succulents is recommended for beginners for they are among the most exotic in appeal, the easiest to propagate, and the most easily grown plants for pots or rockeries where frosts are light. Crown of thorns (*Euphorbia Milii*) Spurge Family: A profuse bloomer, it is almost always with coral-red, circular blossoms and an occasional small, pointed leaf. Bitter Aloe (*Aloe* sp.) Lily Family: This has brilliant orange-red spikes resembling those of poker-plant. Its whorls of succulent leaves are curving and decorative. Flopper (*Kalanchoë Daigremontiana*) Orpine Family: One of the most interesting of leaf propagating plants. Pin a leaf on a curtain and soon roots and tiny plants will appear from its margin. Slipper Plant (*Pedilanthus tithymaloides*) Spurge Family: Its fleshy, zigzag stems bear sparingly pinkish, variegated leaves and small lopsided flowers like miniature red birds which have given it the erroneous name "bird-cactus." It does best in sun, with good drainage, and no frost.

BRAZILIAN PEPPER (*Schinus terebinthifolius*)　　　　Cashew Family

An outstanding shrub or small tree for Christmas decoration in warm climates. Its masses of small red berries make it as effective as holly. Male and female flowers are borne on different plants, so it is necessary to have both for fruiting. Zones 1 and 2; parts of 3. Easily grown from seed or large cutting. Thrives in dry, sunny locations.

ARDISIA (*Ardisia crenulata*)　　　　Myrsine Family

Highly prized for Christmas decoration, for it is covered with showy clusters of brilliant crimson berries from fall till spring. An excellent low, evergreen shrub for full shade. Easily grown from seed or cutting in rich, moist soil in shade. Difficult to transplant. Excellent for pot culture.

POINSETTIA (*Poinsettia pulcherrima*)　　　　Spurge Family

Named for Hon. J. R. Poinsett, U.S. minister to Mexico, who introduced it and exhibited it in Philadelphia in 1829. The Christmas plant supreme in the North is a small tree of tropical America. Even in Fla. with some protection it will reach up to a second-story window. Zones 1 and 2; risky in 3. Prune 2 or 3 times after flowering and again in summer or early September for more and larger flowers. When making cuttings, immerse immediately in boiling water or seal end with paraffin or flame.

AFRICAN TULIP-TREE (*Spathodea campanulata*)　　　　Bignonia Family

A sensational tree from s. Afr., *Spathodea* is popular in s. Fla. and Calif. as a roadside tree, for from winter into summer it bears great clusters of huge tulip-shaped flowers set off by compound leaves with 3-19 lance-shaped leaflets 3"-4" long. Sometimes called "fountain-tree" from the water secreted by its flowers. Grown from cutting or from its small seed. Flowers last long in arrangements. Dropped flowers are useful.

Crown of Thorns, Bitter Aloe,
Flopper, Slipper Plant

Brazilian Pepper and Ardisia

Poinsettia

African Tulip-Tree

Jacaranda

Dwarf Poinciana

Shell Ginger

Flame Vine and Thunbergia

JACARANDA *(Jacaranda mimosaefolia)* Bignonia Family

The most beautiful of all the blue flowering trees. Large, erect racemes bear loose clusters of 40-90 tubular, lavender-blue to mauve blossoms which stand out at the tips of gnarled, grey branches, for they blossom slightly before the new, green, mimosa-like foliage appears. These graceful leaves have a rare, lacy appearance. Seedlings are used in pots. Zone 1 and parts of 2. From seed or cutting. Tree matures before blossoming. After conditioning in water, keeps for several days. Buds open in the house.

DWARF POINCIANA *(Poinciana pulcherrima)* Pea Family

Rarely exceeding 10′ in height, this superb, lacy shrub is topped by racemes of vivid red and yellow flowers with graceful, showy red stamens 2½″ long. Flowering intermittently, its seasons of richest bloom are fall and spring. Zones 1 and 2; risky in 3. Short-lived, but easily propagated by seed soaked overnight. Good for sea coast. Keeps well if picked early or late in the day.

SHELL GINGER *(Alpinia speciosa)* Ginger Family

Native of e. Asia, this tuberous-rooted plant of dense, coarse foliage is a relative of the true ginger of commerce. Blooming in spring and intermittently during the year, its graceful clusters of waxy-white and shell-pink buds nod in drooping racemes half-hidden beneath sheaths of long, dark-green leaves. Close inspection reveals striking markings on the inside of the crinkled lip and a lining of bright yellow, traced with maroon, recalls the tints of certain sea shells, hence the name "shell flower." The blossoms, without the foliage, are effective for arrangement with brightly colored tropical leaves. *Alpinia* grows well in dense shade. Zones 1 and 2. Propagated by division.

FLAME VINE *(Pyrostegia ignea)* Bignonia Family

The Greek words *pyr* and *stege,* meaning *fire* and *roof,* give the botanical name to this spectacular vine from Brazil. Climbing by twining tendrils, the foliage and bloom engulf everything in their path, smothering even the giant pines. As the tubular blossoms age, they dangle for days on the end of the pistil, like a flame in the wind. Zones 1 and 2.

BENGAL CLOCK-VINE *(Thunbergia grandiflora)* Acanthus Family

In warm weather, its heavy green foliage is continually decorated with huge flower clusters of azure. Opening wide from a short tube, the 5-petaled blossoms often attain a spread of 3″. Zones 1 and 2. Cut back after frost. Rejuvenates quickly. Grown from cutting or shoots. Rarely used for cutting as the flowers are prone to wilt.

pale purple

Pycnothymus rigidus

↑
white spotted lavender

Monarda punctata   **Conradina puberula**

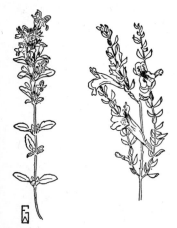

Clinopodium georgianum   C. coccineum

Monarda didyma   red

RED-PENNYROYAL (*Conradina puberula*) Conradinas are small shrubs, with long, slender branches and narrow, entire, revolute leaves in clusters. Flowers are borne in axillary cymes, scattered or grouped. Only 4 species are known and they occur mostly in Fla. *C. grandiflora* with pale-purple, dotted flowers, grows in the lower east coast region, while *C. puberula* with red flowers is restricted to pinelands of the northern Gulf Coast.

BASILS (*Clinopodium*) Basils are shrubs or herbs with broader leaves than pennyroyals (*Pycnothymus*) and with straight corolla tubes. Red-basil (*C. coccineum*) is a slender shrub with bright scarlet flowers blooming nearly all year. The two-lipped corolla has a notched upper lip which is shorter than the triangular, lobed lower. It is frequently found in sandy soil of shores, sandhills, and hammocks, Coastal Plain, Fla. to Ala. and Ga. *C. georgianum,* with white or pink, purple-spotted corollas, grows on riverbanks, in dry woods and hammocks, Coastal Plain, Fla. to Miss. and N.C.

HORSE-MINTS, MONARDAS (*Monarda*) Mostly perennial herbs with strongly 2-lipped corollas and 2 stamens. Flowers in remote or approximate whorls with white, pink, reddish-purple or scarlet corollas. Usually showy and several are grown as ornamentals. Spotted bee-balm (*M. punctata*) is an unusual species with yellowish to cream-colored corollas, spotted with purple. Flowers are subtended by pointed, recurved bracts which are colored purplish-lilac or whitish. White-flowered forms are not infrequent. It is frequently seen near the coast in sandy soil between the dunes but occurs elsewhere in the Coastal Plain and occasionally in other localities. Fla. to Tex., n. to N.Y. Excellent for gardens and distinctive in flower arrangements. Bee-balm, Oswego-tea (*M. didyma*) is by far the showiest and best known of the southern Monardas because of its large scarlet corollas grouped at the summit of the stem. The upper lip of the corolla is long and pointed and the lower somewhat recurved. It grows in moist ground around springs, seepages, and edges of streams. Mostly in upland districts in our area, Ga. to Ala. and northward. The commonly cultivated *M. media* has purplish-red corollas. Several species with pink corollas are equally attractive and should be cultivated also.

[114]

FIGWORT FAMILY (Scrophulariaceae)

Turtle-Heads, Snake-Heads (*Chelone*) Perennials with opposite leaves and attractive, irregular flowers, ranging in color from pure white or greenish to pink or purple. Blooms in autumn.

Balmony (*Chelone glabra*) A highly variable species of several varieties and forms. Corollas are usually white but often pinkish tinted at summit or with greenish-yellow lobes. A purple or deep-rose form is called var. *elatior*. *C. glabra* is distinguished from other species with similar colors by its almost sessile leaves or winged petioles. It is the widest ranging of our species, occurring in swamps and marshy ground in various localities from Ga. n. to e. Can.

Purple Turtle-Head (*Chelone obliqua*) This resembles the preceding, but the leaves are distinctly petioled, lanceolate, and tapering gradually from base to apex. Corollas are rose to magenta, the lower lip with pale yellow beard. A Coastal Plain species, growing in alluvial swamps from Fla. to Ark. and n. to Tenn. and Md. *C. Cuthbertii*, with violet-purple corollas, broadly striped on the anterior side occurs in uplands of N.C. and extends into the Coastal Plain of Va. The purple-flowered *C. Lyonii* which is similar to *C. obliqua* but with broader leaves on long, slender petioles is naturally restricted to borders of mountain streams in N.C. and Tenn. but has escaped from cultivation farther north.

Long-Stalked Monkey-Flower (*Mimulus ringens*) Evidently the famous Swedish botanist, Linnaeus, had a sense of humour, for when he saw the resemblance of this flower to a monkey's face he called it *Mimulus*, meaning little mimic. It is a good example of a "two-lipped" flower, with lips closed although *ringens* means gaping. It is remarkably adapted for cross pollination by insects. It frequents marshy places in various situations with a northern range but extends southward in uplands as far as Ga. The short-stalked monkey-flower (*M. alata*) differs from the above in shorter flower stalks, winged stems, and deeper-colored flowers. It has a more southern distribution, ranging as far south as Fla. and Tex.

[115]

Chelone glabra- white tinged pink

Chelone obliqua- rose to magenta

Mimulus ringens - lavender

purplish   white

Penstemon australis · P. laevigatus

coral *

Penstemon Murrayanus - scarlet

Lav-blue   yellow

Linaria canadensis Gerardia virginica

BEARD-TONGUES (*Penstemon*) These herbaceous perennials are called beard-tongues in allusion to a sterile stamen which is exerted from the corolla as a hairy appendage. A few species with showy corollas are cultivated. The showiest as well as the greatest number of species are in the West. Southern penstemon (*P. australis*), as the specific name indicates, is a southern species which ranges in the Coastal Plain from Fla. to Miss., and s.e. Va. It is distinguished by its bright reddish-purple corollas, and grows in open pinelands and sandy ridges of acid soil. Smooth beard-tongue (*P. laevigatus*) is the most common species of the Piedmont but extends into adjacent areas. It is smooth, or nearly so, and the corolla is light violet-purple to almost white. N. Fla. to Ala., Tenn. and Pa. Red beard-tongue (*P. Murrayanus*) is a southwestern species, restricted to Tex. and Okla., extending into w. La., but is cultivated to some extent further east. A beautiful species with bright-red corollas, blooming from mid-March to mid-June.

BLUE TOAD-FLAX (*Linaria canadensis*) In early spring, this relative of the snapdragon often grows in such abundance in fields, especially of the Coastal Plain, that it gives them a violet hue. A winter annual, it develops a rosette during the winter months, with short branches covered with small, roundish leaves, and, in spring, a slender flowering stem with small flat leaves and a raceme of relatively small, violet flowers with a light "palet" and a slender spur. Although classed as a weed it is highly ornamental and may be used to advantage in flower arrangements. Various localities but varying in abundance, Fla. to Tex., n. to Mass.

YELLOW- or FALSE-FOXGLOVES (*Gerardia*) During the scarcity of wild flowers in upland woods in the southern mid-summer, it is pleasant to see the false-foxgloves with their large, yellow, inflated, trumpet-shaped flowers. Mostly perennials, they are supposed to be partly parasitic on roots of oaks. Sometimes called *Dasystoma*. Downy false-foxglove (*G. virginica*) is less selective as to habitat than most other species. This species grows mostly in open, mixed hardwoods in various provinces from Fla. to La., northward.

[116]

COMB-LEAVED FALSE-FOXGLOVE (*Gerardia pectinata*) This glandular-hairy annual is easily recognized by its finely dissected leaves which are often purplish-tinted. It is a southern species, growing in open, sandy woods of the Coastal Plain and adjacent areas, Fla. to La., n. to Mo., Ky., Tenn. and N.C.

Gerardia pectinata yellow

PURPLE-GERARDIAS (*Gerardia*) The delicate pink flowers of the Gerardias add much to the beauty of autumn flowers. In upland localities, they are usually not seen in great abundance, but in certain habitats in the Coastal Plain they are so numerous that they give a distinctive pink color to open areas. The 2 most frequent species are *G. purpurea* and *G. tenuifolia*. The former has large flowers with corolla lobes hairy at base. Damp, mostly acid soil in various localities, Fla. to Tex., northward. The latter occurs in dry soils, chiefly inland from Ga. to La., northward.

rose-purple   *purple*
Gerardia tenuifolia G. purpurea

BLUE-HEARTS (*Buchnera*) Mostly low, hairy perennials with unbranched stems and opposite leaves which are mostly basal. Flowers borne in slender spikes, axillary to small bracts. Corollas are violet or white with slender tubes, the spreading lobes notched and somewhat irregular. *B. elongata* has rough-pubescent, lanceolate to linear-lanceolate leaves and is restricted to prairies and pinelands of s. Fla. *B. floridana*, with broader leaves, occurs as far north as N.C. *B. americana* with distinctly veined, variously-toothed leaves, is widely distributed in e. U.S. Mostly spring and summer bloomers but in Fla. all year.

INDIAN PAINT-BRUSH, PAINTED-CUP (*Castilleja coccinea*) The common name paint-brush is very suitable for this remarkable plant as it seems to have been dipped in red paint. The showiest part is not the two-lipped, yellow-greenish corolla but the subtending bracts which are scarlet in the upper portion. It grows in meadows and low, open woods, and where in great abundance gives a reddish color to vegetation. Usually in acid soil, from the Coastal Plain to the mountains, widely distributed in e. N.A.

Buchnera elongata Castilleja coccinea

bluish

Dyschoriste oblongifolia. Ruellia carolinensis

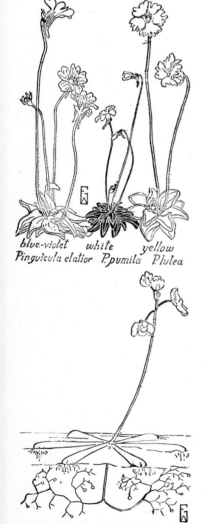

blue-violet   *white*   *yellow*
Pinguicula elatior   *P. pumila*   *P. lutea*

Utricularia inflata - yellow

ACANTHUS FAMILY (Acanthaceae)

CALOPHANES (*Dyschoriste oblongifolia*) Among the relatively rare blue flowers of dry pinelands and sandy ridges of the Coastal Plain, those with a short tube and bell-shaped throat belong to calophanes. Of the 3 species, *D. oblongifolia* has the largest flowers, nearly 1" long, grouped at the tip of the stem. Coastal Plain, Fla. to Va.

RUELLIAS (*Ruellia carolinensis*) Similar to the above, but flowers with a longer tube and not distinctly grouped but in the axils of the upper leaves. Attractive, but flowers fall as soon as picked. Dry soil, Fla. to Kans., Mich. and N.J.

BLADDERWORT FAMILY (Lentibulariaceae)

BUTTERWORTS (*Pinguicula*) Interesting small insectivorous plants with a rosette of flat, pale-green leaves, incurved at the edges, sending up in spring a slender, naked stalk bearing a single flower. Corolla united, irregular, with a spur, the lobes variously notched. Two stamens inserted inside the corolla tube. *P. caerulea* has a delicately pale-violet corolla. Fairly frequent in low pinelands, Fla. to N.C. *P. lutea* is very similar to the above but corollas are a golden-yellow. Similar habitats and distribution, except the latter extends to La. *P. pumila,* as the name suggests, is small and has a white flower. Habitat and distribution the same as the yellow butterwort but does not extend as far north. *P. planifolia,* with a sac-like spur and violet corolla, occurs in shallow water along the Gulf Coast.

BLADDERWORTS (*Utricularia*) On margins of ponds and swamps, one often sees small, yellow or purple flowers on slender stalks extending above the surface of the water or from mud or moist sand. These are bladderworts, so-called because of the small sacs borne on their dissected, thread-like leaves, which by a trap-door mechanism catch small aquatic animals. *U. subulata* grows in wet, sandy soil, has few delicate leaves at base which soon disappear. Coastal Plain, Fla. to Tex., northward. *U. inflata* floats by means of a conspicuous rosette of flat, inflated floats. Flowers are yellow with a prominent 2-lobed palate. Coastal Plain, Fla. to Del. *U. cornuta* is another frequent species on borders of ponds throughout our area. It is distinguished by relatively large, yellow flowers in a spike, each with 2 bractlets at base.

[118]

TRUMPET-CREEPER FAMILY (Bignoniaceae)

A family of trees, shrubs, and woody vines (rarely herbs). Flowers large, trumpet-shaped, and mostly in shades of yellow.

TRUMPET-CREEPER (*Campsis radicans*) The most typical woody vine of the South, often seen climbing telephone poles along the highways, tumble-down cabins and fences and covering waste places. Leaves are pinnately compound; flowers trumpet-shaped, ranging in color from pale yellow to orange or red. Attractive but difficult to eradicate because of its deep roots which sprout when broken off. Also called "cow-itch-vine" because of its reputed poisonous qualities to the skin. Reported to cause dermatitis in susceptible individuals. In various provinces, Fla. to Tex., n. to Mo. and N.J. The long pods are often used in arrangements.

Campsis radicans - red to orange

CROSS-VINE (*Bignonia capreolata*) Called cross-vine in allusion to the image of a Greek cross seen in the cross section of its stem. Leaves 2-foliate compound, usually with tendrils. Corolla arched trumpet-shaped with somewhat irregular, spreading lobes, red-orange without, yellow within. An attractive native vine. River bottoms and low woodlands, Coastal Plain, and low Piedmont, Fla. to La., n. to Ill. and Va. Easily grown from seed. Makes attractive bowers.

stem X

Bignonia capreolata - red-orange

MISTLETOE FAMILY (Loranthaceae)

MISTLETOE (*Phoradendron flavescens*) Soft-woody, green plants with opposite leaves, parasitic on various deciduous trees, especially black-gum, maples, and oaks. The mistletoe of Europe was considered of panacean and magic virtues by the Celtic druids. This has been carried over into the custom of hanging up its white berries at Christmas season as a challenge to the young man to kiss the girl who intentionally or unintentionally stands under them. Most frequent in the Coastal Plain but also in other areas, Fla. to Tex., n. to Okla. and N.J.

Phoradendron - berries whitish

Lilac to white
Houstonia purpurea

Lilac, white or
H. caerulea - blue

MADDER FAMILY (Rubiaceae)

BLUETS, INNOCENCE (*Houstonia*) The cerulean bluets are among the first flowers to bloom in spring and are, therefore, welcomed as little friends who have returned after the cold and seemingly lifeless winter season. The first to appear is the small, winter-annual *H. patens* with a deep-purple corolla and dark center. Sometimes called "small-bluets" or "star-violets." Common in fields, Fla. to Tex., n. to Ill. and Va. Next comes the pale-bluish *H. caerulea* with a yellow center, a perennial of open hardwood slopes and stream bluffs. Both of the above may have white flowers. In mountain springs, is the creeping bluet (*H. serpyllifolia*) with flowers similar to *H. caerulea* but with long slender stems and small, roundish leaves. Another group of Houstonias are larger plants with larger leaves and flowers grouped at the summit of the stem. An example of these is purple houstonia (*H. purpurea*), widely distributed in the South.

Randia aculeata - white

RANDIA (*Randia aculeata*) An evergreen, usually spiny shrub with relatively few leathery leaves with wedge-shaped blades. Small, delicate white flowers are borne in the leaf axils. The fruit is an oval or almost spherical greenish-white berry. Hammocks and pinelands, s. Fla. and the Keys. A distinctive and beautiful shrub which may be cultivated as an ornamental.

BUTTONBUSH (*Cephalanthus occidentalis*) A widely distributed shrub growing in wet places, such as margins of ponds and streams, in marshes and swamps. The delicate white, tubular flowers with 5-lobed corollas and conspicuously protruding styles are grouped in a spherical head called "button" as in button-wood or sycamore. Associated with cat-tails, lizard's-tails, sedges, and rushes. Visited constantly by butterflies and bumblebees. Various districts, Fla. to Tex., n. to Can.; also in Calif.

[120]

Cephalanthus occidentalis   white

HAMELIA (*Hamelia patens*) Often called "fire-bush" because of its reddish inflorescence. The crimson corollas are fluted, narrowly tubular, with 5 short lobes. The calyces as well as the branches of the inflorescences are also reddish in color. Conspicuous on edges of hammocks in s. Fla. and the Keys; also in the W.I.

Hamelia patens - red-orange-black berry

WILD-COFFEE (*Psychotria undata*) The Psychotrias are mostly shrubs with opposite, thinnish leaf-blades and small flowers in terminal cymes. Corollas are white or green with 5 strongly reflexed lobes. The fruit is a small, 2-seeded, crimson berry-like drupe as in true coffee, which, however, may vary to white or yellow. *P. Sulzneri* has greenish corollas and long scarlet, orange, or yellow fruits. Both occur in hammocks and woods of s. Fla. and the Keys; also in W.I.

Psychotria undata - white - crimson-drupe

SNOW-BERRY (*Chiococca alba*) This attractive shrub should perhaps be called the "southern snow-berry" since another "snow-berry" (*Symphoricarpos albus*) occurs in the North. The former is a large, erect, diffuse, or reclining plant with elliptic or oval leaves. Small white flowers are borne in axillary, raceme-like cymes. The corolla tube is inflated trumpet-shaped with 5 reflexed lobes. Fruit is a white, berry-like drupe. The natives of the Florida Keys, not knowing much about snow, call it "rat-root" from the ratty odor of its roots. Hammocks, s. Fla. and the Keys.

[121]

Chiococca alba - white

Mitchélla repens white

Sambucus Simpsonii - white

1Viburnum obovatum 2V. nudum

TWIN- or PARTRIDGE-BERRY (*Mitchella repens*) A trailing, evergreen undershrub, with white corollas, densely white-hairy within. Unique in that flowers are borne in pairs with partly joined ovaries which completely merge at maturity into a red berry-like fruit consisting of twin drupes. Damp woods and borders of swamps and bogs, usually in acid soil, in various areas but most abundant in the Coastal Plain. Fla. to Tex. and northward. Used extensively with mosses and lichens as an ornamental in closed glass bowls in which they will keep for months. The only other species known occurs in Alaska.

HONEYSUCKLE FAMILY (Caprifoliaceae)

Mostly shrubs, trees, and vines with opposite leaves, united petals, and fruits usually fleshy berries or drupes.

ELDERS or ELDER-BERRIES (*Sambucus*) The elderberries are mostly shrubs or sub-shrubs, with large pinnately compound leaves, the leaflets serrate, and numerous small flowers in large flat-topped cymes. The fruits are small dark-purple or red drupes commonly called "berries." Three species occur in the South: *S. canadensis,* or common elder, with purple fruits, has a northern distribution but ranges as far south as Ga. and Tex; *S. Simpsonii,* gulf- or southern-elder, also with purple fruits, occurs from Fla. to La.; red-berried or mountain-elder (*S. pubens*) occurs at higher altitudes as far south as Ga.

VIBURNUMS (*Viburnum*) Like elder-berries, the Viburnums have also flat-topped inflorescences but differ in having simple leaves and are definitely woody, being either shrubs or small trees. Southern black-haw (*V. rufidulum*) is a small tree with attractive lustrous, elliptic or obovate leaf-blades and rusty petioles and buds. The drupes are deep blue. It blooms about 2 weeks later than the northern black-haw. Woods and thickets, Fla. to Tex., n. to Ill. and Va. Northern black-haw (*V. prunifolium*) is a shrub or small tree with relatively small, oval or ovate, finely serrate leaves, the petioles and buds not rusty scurvy. Drupes, blue-black. Habitats and distribution similar to the preceding. Both are of ornamental value and should be used as ornamentals more than they are, especially the southern black-haw.

[122]

JAPANESE-HONEYSUCKLE (*Lonicera japonica*) Like many plants which have migrated into our country, this woody vine has weedy characteristics, flourishing to an extent which is detrimental to our native vegetation. But in spite of our prejudice against it, we admit that its flowers are beautiful and their fragrance delightful. The irregular corollas are white when young, fading to yellow with age. Berries, black. Although injurious in some places, it no doubt serves a useful purpose in others in reducing erosion. Especially luxuriant in river bottoms. Throughout the South, extending into some of the northern states. A form with purplish coloration has been named var. *chinensis.*

Lonicera japonica - white to yellow

TRUMPET-HONEYSUCKLES (*Lonicera*) One of the most ornamental of our woody vines is the scarlet-flowered trumpet- or coral-honeysuckle, also called woodbine (*L. sempervirens*). The fruits are red berries. In various situations from Tex. and Fla. northward. Widely cultivated and escaping beyond its natural range. Yellow-honeysuckle or woodbine (*L. flava*) is rather infrequent with orange-yellow flowers similar in shape to the japanese-honeysuckle. Grows in rocky woods from Ga. to Ala., Tenn., and N.C. A very attractive vine which is often used as an ornamental.

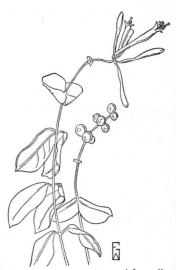

Lonicera sempervirens - scarlet & yellow

BIRTHROOT or DUTCHMAN'S-PIPE FAMILY
(Aristolochiaceae)

WILD-GINGER (*Asarum canadense*) A trailing, perennial herb with cordate or kidney-shaped, deciduous, hairy leaves and short-stalked, apetalous flowers with a 3-lobed, brownish-purple calyx. The plants are pleasantly aromatic when bruised. Variable, especially in its calyx lobes, and, therefore, segregated into 3 or 4 varieties considered by some as species. The one illustrated is var. *reflexum* with triangular, reflexed calyx lobes. Another variety occurring in our area is var. *acuminatum* with calyx lobes drawn out into long, slender tips. In wooded slopes from N.C. northward.

[123]

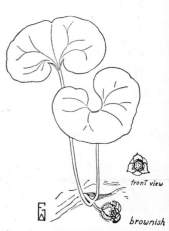

front view

brownish

Asarum canadense var. reflexum

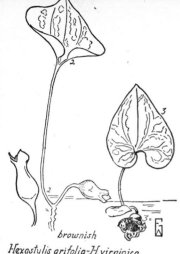

HEART-LEAFS (*Hexastylis*) These low, perennial herbs are closely related to wild-ginger and sometimes placed in that genus. They are also aromatic, but differ from the latter in being smooth and evergreen, without trailing stems, and also different flower characters. The leaves vary in different species from arrow-shaped to heart-shaped with pointed or rounded tips. Flowers vary from vase-shaped (*H. arifolia*) to bell-shaped (*H. virginica*). The former is the most widely distributed, occurring in various localities from Fla. to Va. The latter ranges from Va. to Ga. Several other species occur in the South. Because of their evergreen leaves, which are often mottled light-green, they are attractive plants in wild-flower gardens.

brownish

Hexastylis arifolia-H.virginica

GOURD FAMILY (Cucurbitaceae)

A family of annual or perennial, tendril-bearing vines, with unisexual flowers, including such well-known plants as cucumber, squash, pumpkin, and watermelon.

WILD BALSAM-APPLE (*Momordica Charantia*) A creeping or climbing herbaceous vine with 5- to 7-lobed leaves, bearing both staminate and pistillate flowers on the same plant (monoecious), both with yellow corollas. A curious-looking bract is borne on the staminate flower. The female flower has a slender calyx-tube above the inferior ovary. The fruit is a pointed golden-yellow, spiny berry, 3″-4″ long. In thickets and waste places, outer Coastal Plain, Fla. to Tex.

Momordica Charantia-yellow-seed red

BELL-FLOWER FAMILY (Campanulaceae)

MOUNTAIN-BLUEBELL (*Campanula divaricata*) Bluebells or bellflowers are more abundant in northern latitudes, so it is not surprising to find the only representatives we have to be mostly restricted to higher altitudes. The most delightful of these is the mountain-bluebell with many, drooping, small, blue flowers hanging over ledges and road-cuts in seepages, on edges of mountain streams and waterfalls. Ga. to Ala., Ky., W.Va., and Md. In the American-bluebell (*C. americana*), the corollas are deeply lobed, the lobes spreading or reflexed, differing so much from typical bellflowers that it is sometimes placed in another genus (*Campanulastrum*). Mostly at higher altitudes, Fla. to Ark., northward.

[124]

Campanula divaricata - blue

VENUS'S LOOKING-GLASS (*Specularia perfoliata*)
The attractive purplish flower is "Venus" and
the leaf, or bract, at its base is the "looking-
glass." It is a common annual of fields, open
woods, and waste ground, blooming in spring.
The flowers are usually borne only on the upper
portion of the stem, but, occasionally, there is a
flower in the axil of every leaf, the lowest of
which are self-pollinated (cleistogamous). The
leaves are clasping, but in the closely related
species, *S. biflora,* they are not. The slender seed
pod is interesting in having 3 openings on the
sides for the shedding of seed. Ranges from Fla.
to n. Mex. and n. to the northern states.

Specularia perfoliata violet-blue

LOBELIA FAMILY (Lobeliaceae)

This family, mostly herbs, was named for a noted
Flemish herbalist, Matthias de l'Obel (1538-
1616). The only genus represented in our area
is *Lobelia* of about 20 species.

SWAMP-LOBELIA (*Lobelia glandulosa*) This tall
(up to 3') slender-stemmed species with its deep
blue or lavender corolla with a white center
stands out conspicuously in contrast to the usually drab surroundings of grasses, sedges, and
rushes which make up most of the vegetation of
extensive, wet marshlands of our lower Coastal
Plain. The glands at the tips of the marginal
teeth of the long narrow leaves give it its specific
name. It ranges from Fla. to Va.

Lavender
Lobelia glandulosa

CARDINAL-FLOWER (*Lobelia cardinalis*) The brilliant red of this flower gives one a thrill as it
flashes from the marshlands. This is an exception to most species of *Lobelia* which are in
shades of blue, variegated with white, or all
white. It has the largest flower of all our Lobelias. Grows in wet ground or borders of
streams, marshes, and swamps in various localities from Fla. to Tex. and northward. Easily
grown from seed in moist soil.

[125]

red
Lobelia cardinalis

GREAT-LOBELIA (*Lobelia siphilitica*) This is the largest of our blue Lobelias and is sometimes called by the paradoxical name "blue cardinal-flower." The flowers are crowded at the end of a smooth to slightly hairy stem bearing relatively large, broadly lanceolate, irregularly toothed leaves. The sepals are strongly auricled at base, and the capsule is hemispherical and short-beaked. Widely distributed in low grounds, meadows, and borders of streams, Ga. to La. and northward. Best cultivated from seed.

Lobelia siphilitica - bluish

GOODENIA FAMILY (Goodeniaceae)

SCAEVOLA (*Scaevola Plumieri*) A curious plant, closely related to Lobelias, and the only representative of this family in the s. U.S. The pink or white flowers are relatively large with a very irregular corolla, woolly within, split on one side so that the pistil and stamens protrude conspicuously. The prominent lobes are crisp-winged on the margins. The plant is a succulent, growing on coastal dunes where the sprawling stems often spread under the sand, forming extensive colonies. The leaves are mostly wedge-shaped with smooth edges. The fruit is a black, juicy berry. S. Fla., the Keys, and W.I.

Scaevola Plumieri - white

COMPOSITE FAMILY (Compositae)

STOKESIA (*Stokesia laevis*) This attractive perennial is so widely cultivated that it is not generally recognized as a native of the South. A unique plant, for only one species is known. The large blue, lilac, or white heads are different from most composites in that the flowers vary gradually from the large outer flowers with flat corollas to the small inner blossoms with tubular corollas. It is native to moist pinelands of the Coastal Plain from Fla. to La. and S.C. but is cultivated as far north as N.C.

Stokesia laevis purplish-blue

COMPOSITAE

The Compositae is the largest and most widely distributed of all families of plants. Its most outstanding characteristic is the inflorescence which consists of an aggregate of usually many relatively small flowers arranged in a "head" so that the group simulates a single flower. The flowers are surrounded at base by a number of bracts, the involucre, which might be mistaken for a calyx; when ray flowers are present, as in Asters, dahlias and chrysanthemums, they may be taken for petals of the "flower" instead of corollas of individual flowers. The name Compositae, or composite, means "compound" which was the concept of this type of a "flower" of ancient botanists. In ordinary parlance, the head is called the "flower" while the individual flowers of which it is composed are called "florets." Although the individual flowers are constructed on a similar plan, they vary considerably, especially in shape and color of corollas as well as in function. The variation in the appearance of heads depends mainly upon the types of florets present and their arrangement. The florets are in general of 2 kinds, those with short, funnel-shaped, or tubular, corollas and those with longer strap-shaped ones. In some plants, such as tansy, the flowers are similar and tubular, or funnel-shaped. In others, such as dandelion, all flowers are strap-shaped. The latter type makes up a distinct group of composites usually with milky juice, or latex, which in the Russian dandelion is a source of rubber. In the majority of composites, however, the florets of the head are of 2 kinds, those occupying the center are small, tubular, and the outer or marginal, are strap-shaped. The former are commonly called "disk flowers," and the latter, "ray" or "ligulate." The 2 kinds may have corollas of the same color as in golden-asters, or of different color as in Asters. The outer ray flowers may be pistillate and bear fruit, or neutral, functioning only for show, but never staminate. The central are always perfect but sometimes function only as staminate. The fruits of composites are dry and one-seeded, called botanically "achenes," but by most people are simply called "seeds." In the majority of composites, the achenes are provided on top with hairy bristles, barbed spines, or scales which are modified calyces adapted for distribution by wind or animals. Most composites are perennial herbs, but a few are woody, such as the shrubby groundsel-tree (*Baccharis*) of our coastal areas, but in some parts of the world they may attain small tree size. Being showy and ornamental plants, the composite family has contributed a large number of cultivated flowers, such as sun-flowers, asters, zinnias, marigolds, dahlias and chrysanthemums. In many varieties of these, the central flowers have been transformed into ray flowers and the head is then said to be "double." Most of the garden composites are autumn-blooming as are those in nature. But, curiously enough, the Compositae have not contributed proportionately as many to our cultivated edible plants. However, a few may be mentioned, such as lettuce, salsify, bur-artichoke (the soft receptacle and the base of the involucral bracts are eaten), and the Jerusalem-artichoke, or girasole, a species of sun-flower noted for its edible tubers.

MIMOSA or SILK-TREE *(Albizia Julibrissin)*　　　　　　　　Pea Family

One of the South's outstanding ornamental flowering trees for road-side planting is the picturesque, flat-topped, low-spreading (20'-40') mimosa tree, with foliage of graceful, feathery, doubly compound leaves which characterize so many members of this large family. A native of Persia and Japan. Ethereal brushes of clusters of numerous slender stamens cover the tree in a soft haze of pink from late spring throughout most of the summer. Hardy in Zones 3 and 4 and in the mountains n. to Washington, D.C. Easily and quickly grown from seed but short-lived. Unfortunately now being attacked by a blight which is killing out many trees in certain sections. Too fugacious for picking.

WHITE-RHODODENDRON *(Rhododendron maximum)*　　　　Heath Family

Great-laurel or rose-bay is the most common rhododendron seen in the s. Appalachian Mountains during early summer. It has thick, evergreen leaves without glandular dots. Corollas are mainly white or pinkish-tinted, the upper lobe with green dots at the base. Pedicels are viscid. Calyx lobes, longer than broad. Sometimes forming thickets on mountain slopes called locally "woolly heads," but frequenting stream banks, rarely extending into the Coastal Plain. N. Ga. to Ala. northward.

GEORGIA BARK or FEVER TREE *(Pinckneya pubens)*　　　Madder Family

This is another example in which showiness of flowers and inflorescences is not restricted to petals, for the colorfulness of Georgia bark is located in its sepals, some of which resemble bright-pink leaves. The corolla is greenish, mottled with brown or purple. A very unusual and attractive shrub. Sandy swamps, Coastal Plain, w. Fla. to S.C.

CAMELLIA JAPONICA VARIETIES　　　　　　　　　　　　　Tea Family

For hundreds of years the stately formality of the symmetrical *Camellia japonica* has been the pride of southern gardens. This aristocrat of the famous tea family originated in China, but the recent camellia popularity has caused the development of hundreds of new varieties and fancy double hybrids. The waxen blossoms are striking against the glossy, dark-green leathery foliage. Slow growing, but many ancient bushes in the Charleston gardens are 25'-30' tall. A careful selection of varieties results in a succession of bloom from Oct. into May. Zones 1-3; Piedmont and Oregon. Varieties vary in hardiness; like partial shade, well-drained, acid soil, rich in humus and manure; no lime. Propagate from short cutting taken in winter, set in pure, white sand. Use sparingly in arranging. Cut blossoms carefully instead of pruning to shape plant.

Mimosa or Silk-Tree

Rhododendron

Georgia Bark or Fever Tree

Camellia japonica varieties

FALL WILD FLOWERS OF THE SOUTHERN MOUNTAINS

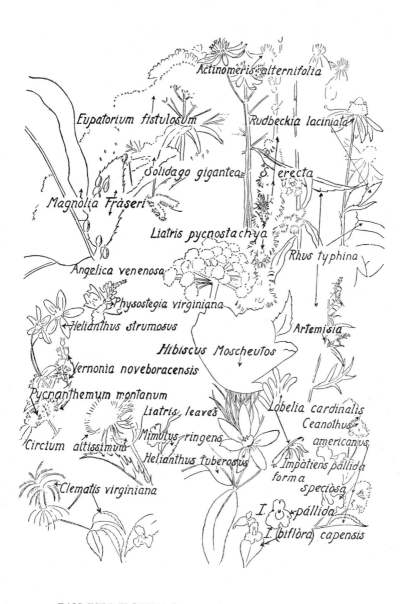

FALL WILD FLOWERS OF THE SOUTHERN MOUNTAINS

IRON-WEEDS (*Vernonia*) These tall perennials with corymbose cymes of bright-purple heads of tubular flowers with conspicuous stigmas deserve a more appropriate common name. The New York iron-weed (*V. noveboracensis*), which grows mostly in marshy ground and has a northern range, is the best known of our species. *V. glauca* is somewhat similar to this but with different-shaped leaves, whitish on the underside, and grows on wooded slopes from Ga. to Ala., n. to N.J. *V. angustifolia,* with long narrow, linear leaves, is the most common species in the Coastal Plain. Sandy woods and pinelands, Fla. to Miss. and N.C. All flower in late summer.

purple

Vernonia noveboracensis V. angustifolia

DOG-FENNEL (*Eupatorium capillifolium*) When the heat of summer changes to the cool, refreshing days of autumn, you may wonder what the tall, bushy plant with feathery leaves is that grows on roadsides, edges of fields, and in overgrazed pastures. The flower-heads are small but so numerous that when in bloom they give the plant a whitish tint. Although classed as a "weed," it is one of the picturesque plants of autumn. Various situations and localities throughout the South. Used in flower arrangements for a feathery effect. There are many other species of *Eupatorium,* the tallest and showiest of which are the joe-pye-weeds of marshy grounds, with large corymbs of purplish heads. *E. aromaticum* is a white-flowered species related to white snakeroot (*E. rugosum*) which causes "milk sickness" of cattle and humans. The beautiful bluish-violet mist-flower (*Conoclinium coelestinum*) is often cultivated and mistaken for ageratum with which it is closely related.

white

Eupatorium capillifolium E. aromaticum

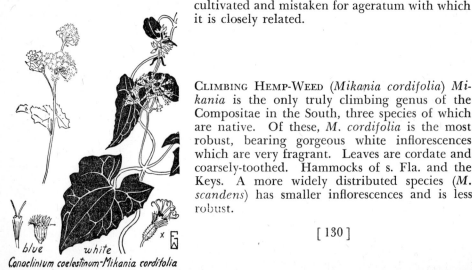

CLIMBING HEMP-WEED (*Mikania cordifolia*) *Mikania* is the only truly climbing genus of the Compositae in the South, three species of which are native. Of these, *M. cordifolia* is the most robust, bearing gorgeous white inflorescences which are very fragrant. Leaves are cordate and coarsely-toothed. Hammocks of s. Fla. and the Keys. A more widely distributed species (*M. scandens*) has smaller inflorescences and is less robust.

[130]

blue   *white*

Conoclinium coelestinum-Mikania cordifolia

CARPHEPHORUS (*Carphephorus*) These simple-stemmed perennial herbs are sometimes difficult to distinguish from other genera with similar inflorescences. The heads are composed of similar, rose-purple, tubular flowers, surrounded by an involucre of scarious bracts. Heads are in open or cluster-like corymbose cymes. In *C. corymbosus,* the heads are clustered and the stem is sparingly hairy and conspicuously leafy up to the inflorescence, while in *C. tomentosus,* inflorescence is more open and the stem is less leafy above and copiously hairy. Both grow in open pinelands in the Coastal Plain, Fla. to N.C.

rose-purple

Carphephorus corymbosus-C. tomentosus

TRILISA (*Trilisa paniculata*) A sticky-hairy, perennial herb, with a narrow-panicle of relatively small, few-flowered, rose-purple heads, blooming in fall and winter. The basal leaves are linear-lanceolate, abruptly reduced upward. At a distance resembling carphephorus but quite different in its inflorescence. Growing in low pinelands, Coastal Plain, Fla. to La. and N.C. Deer's-tongue (*Trilisa odoratissima*) is valued for its abundance of coumarin, a vanilla-like flavoring agent.

GARBERIA (*Garberia fruticosa*) This is one of the shrubby composites inhabiting s.e. U.S. It is easily recognized by its blunt alternate, thickish, evergreen leaves and rose-purple heads arranged in a somewhat dense, corymbose cyme; also when past flowering, identified by its conspicuous brown pappus. The anthers are conspicuously exerted. Sand-scrub and dunes, peninsular Fla.

rose-purple X

Garberia fruticosa-Trilisa paniculata

CAMPHOR-PLANT (*Heterotheca subaxillaris*) The name *Heterotheca* means "different cups" in allusion to its variable pappus which in the inner or disk flowers consists of many long, slender bristles seen at the top of the mature achenes ("seeds"), while in the outer or ray flowers the bristles are shorter and stouter. Heads are relatively large with both the inner, tubular flowers and the outer ray flowers a golden-yellow. A weedy plant, hence found in various situations but northward mostly near the coast in sandy soil. Fla. to Ariz. and n. to Del.

[131]

Heterotheca subaxillaris - yellow

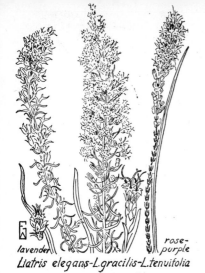

lavender   rose-purple

Liatris elegans-L.gracilis-L.tenuifolia

basal leaf

one flower

Liatris squarrosa L. aspera- purple

yellow

Chrysopsis microcephala C.Tracyi

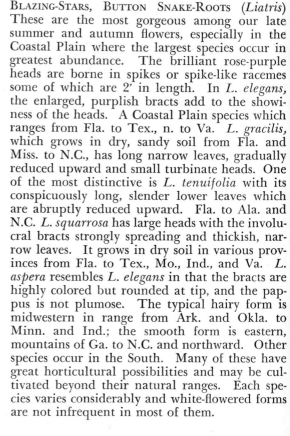

BLAZING-STARS, BUTTON SNAKE-ROOTS (*Liatris*)
These are the most gorgeous among our late
summer and autumn flowers, especially in the
Coastal Plain where the largest species occur in
greatest abundance. The brilliant rose-purple
heads are borne in spikes or spike-like racemes
some of which are 2′ in length. In *L. elegans*,
the enlarged, purplish bracts add to the showiness of the heads. A Coastal Plain species which
ranges from Fla. to Tex., n. to Va. *L. gracilis*,
which grows in dry, sandy soil from Fla. and
Miss. to N.C., has long narrow leaves, gradually
reduced upward and small turbinate heads. One
of the most distinctive is *L. tenuifolia* with its
conspicuously long, slender lower leaves which
are abruptly reduced upward. Fla. to Ala. and
N.C. *L. squarrosa* has large heads with the involucral bracts strongly spreading and thickish, narrow leaves. It grows in dry soil in various provinces from Fla. to Tex., Mo., Ind., and Va. *L.
aspera* resembles *L. elegans* in that the bracts are
highly colored but rounded at tip, and the pappus is not plumose. The typical hairy form is
midwestern in range from Ark. and Okla. to
Minn. and Ind.; the smooth form is eastern,
mountains of Ga. to N.C. and northward. Other
species occur in the South. Many of these have
great horticultural possibilities and may be cultivated beyond their natural ranges. Each species varies considerably and white-flowered forms
are not infrequent in most of them.

GOLDEN-ASTERS (*Chrysopsis*) Much of the yellow
in our autumn flowers is contributed by the
golden-asters. They commence blooming in the
late summer and continue into the late fall.
They have relatively large heads with the central
and ray flowers of the same color. Over a dozen
species have been reported from the southern
states, but their classification is in such an un-
satisfactory state that just how many distinct
species they really represent awaits further study.
C. mariana, with broad leaves and large flowers
is the most common species in upland districts
throughout e. U.S. Most of the others have
longer and narrower leaves which are covered,

like the stem, with silky, silvery- or lead-colored cobwebby-hairs. *C. microcephala,* meaning small-headed, has long, narrow leaves and relatively small heads. It grows in sandy soil, Coastal Plain, Fla. to Tex. and S.C. It is closely related to *C. graminifolia* (grass-leaved) with longer heads. *C. Tracyi* also has narrow leaves similar to the above two but has fewer and larger heads. It is the southernmost species, restricted to s. Fla. and the Keys.

silvery
Chrysopsis graminifolia - C. mariana - yellow

GROUNDSELL-TREE (*Baccharis halimifolia*) This is a shrubby composite in which the flowers are staminate and pistillate on different plants. The pistillate flowers have a long silky pappus on the top of the fruits, which makes the mature inflorescences of ornamental value in arrangements and dried bouquets. It grows in fresh to brackish marshland, mostly near the coast, but is occasionally found inland. Fla. to Tex., n. to Mass. Three other species occur in the South.

MARSH-FLEABANES (*Pluchea*) The small heads crowded at the upper parts of the plant are without ray flowers and look like small brushes. The flowers are all tubular but vary in shape from the outside inward, and in color from creamy to purplish. Most species emit unpleasant odors. *P. foetida* is so called because of its repulsive odor. It has somewhat clasping leaves and creamy flowers. Marshy or swampy situations, Coastal Plain and adjacent areas, Fla. to Tex., n. to N.J. A pink or purplish species with sessile, fleshy leaves and growing in brackish habitats, is the salt-marsh fleabane (*P. purpurascens* var. *succulenta*), ranging from Fla. to Me.

lavender   white
Pluchea foetida - Baccharis halimifolia

BLACK-ROOT (*Pterocaulon undulatum*) A unique and picturesque plant with its white-woolly underside of leaves and winged stems with the decurrent, wavy leaf-bases. The small creamy heads are crowded in spicate-cymes. Related to everlastings, pussy-toes, and rabbit-tobaccos or cud-weeds. Conspicuous in savannas in late spring, Fla. to Miss. and N.C.

Pterocaulon undulatum

GOLDENRODS (*Solidago*) The blooming of golden rods is associated with late summer and autumn. They have numerous small, usually yellow, heads in compound inflorescences. Some 60 species occur in the South. Few are cultivated. Seaside goldenrod (*S. sempervirens*) is one of our most beautiful species and distinctive both in appearance and habitat, for no other goldenrod grows in beach sand and in brackish soil. Plants are robust with long, thickish, somewhat fleshy basal leaves and narrow, compact flower groups. The southern form is var. *mexicana*. Fla. to Tex. and n. to N. Eng.; also in W.I.

yellow

Solidago sempervirens var. *mexicana* S *odora*

SWEET- or FRAGRANT-GOLDENROD (*S. odora*) Another easily recognized species by its smooth edged leaves which are transluscent-spotted and give off a strong odor of anise when bruised. Dry, sandy or clayey, acid soil. Coastal Plain and adjacent areas, Fla. to Tex. and northward. Blooms in late summer. Chapman's-goldenrod (*S. Chapmanii*) is closely related to the preceding having entire leaves with translucent spots. It is also fragrant. However, the stems are distinctly hairy and the leaves oblong, ovate to lanceolate. Confined to pinelands and sand hammocks of Fla. and s. Ga.

yellow

Solidago Chapmanii S. *bicolor whitish*

WHITE-GOLDENROD, SILVER-ROD (*Solidago bicolor*) The only goldenrod in our area with whitish or cream-colored flowers. A very distinctive and beautiful species. The flowers are in a spike-like panicle. Often seen in dry, open woods and rocky stream bluffs. Widely distributed, Ga. to Mo., and northward. Has excellent horticultural possibilities.

GIANT-GOLDENROD (*Solidago gigantea*) This is the tallest of our goldenrods, spreading by underground stems and forming colonies. The lower leaves are prominently 3-nerved. Moist thickets in various situations and widely distributed in N.A. Small-headed goldenrod (*Solidago microcephala*) represents a few goldenrods with flat topped inflorescences which are sometimes placed in the genus *Euthamia*. A Coastal Plain species which ranges from Fla. to La., n. to N.J.

flower

cluster

(*Euthamia minor*)

Solidago microcephala -. S *gigantea*

WHITE-TOPPED ASTERS (*Sericocarpus*) Often associated in habitats and time of flowering with golden-asters. Closely related to goldenrods but looking more like Asters. Differing from the latter in having both center and ray flowers of the same color (white). Four species occur in the South. *S. bifoliatus* is a southern species with hairy, smooth-edged leaves and blunt, firm involucral bracts. Dry pinelands, Coastal Plain, Fla. to La. and Va. *S. linifolius* has uniformly narrow leaves and the widely distributed *S. asteroides* has broad, toothed leaves.

ASTERS (*Aster*) Asters belong to one of the largest tribes of the Compositae, which includes goldenrods, golden-asters, white-topped asters, fleabanes, and others. The central or disk flowers are always yellow while the outer or ray-flowers are of a different color, often purple, blue, pink, or white. The heads are always on leafy stalks. Elliott's-aster (*A. Elliottii*) is one of the tall Asters, growing in swamps of the Coastal Plain from Fla. to N.C. Medium-sized heads are numerous and crowded at the ends of the branches, with bright-purple ray flowers. The stems are smooth or slightly hairy in lines. Sometimes growing in great abundance, Coastal Plain, Fla. to N.C. *Aster concolor* is another aster of dry, sandy open woods, with an unbranched stem and densely, fine-hairy foliage giving it a somewhat grayish-green color. A distinctive feature of this species is its spike-like racemic inflorescence and lilac ray flowers, fading violet-blue. Coastal Plain and lower Piedmont, Fla. to La., n. to Mass.

SPREADING ASTER (*Aster patens*) One of the most delightful Asters of upland hardwoods, with relatively large heads and bluish-purple to violet rays. Easily recognized by its widely spreading branches, stiff hairiness, and clasping leaf bases. Widely distributed in e. N.A. Old-field aster (*A. pilosus*) is a white-flowered aster flourishing in autumn along roadsides and in waste places but especially in abandoned fields where it may grow in pure stands. It is quite ornamental especially when growing in rich ground, but because of its weedy nature it is not often admired. It is usually distinctly hairy but varies in this as well as in flower color which may have a pinkish tint. Widely distributed from Ga. and Miss. northward.

white
Sericocarpus bifoliatus

light purple
Aster Elliottii

violet-purple
Aster concolor

blue-violet
Aster patens

A. exilis - purplish

white
Aster pilosus

[135]

LARGE-FLOWERED ASTER (*Aster grandiflorus*) The largest and the showiest of our southern, native, purple Asters. The large heads, 2″ across, with deep-violet ray flowers terminate long branches. Stems, leaves, and involucral bracts are copiously glandular-hairy. The short, blunt and narrow leaves are somewhat clasping at base. Although the flowers are gorgeous, the plant as a whole is twiggy and, therefore, does not have an especially pleasing appearance. The flowers wilt rather easily when picked. Dry, sandy, or clayey soil, Coastal Plain and lower Piedmont, Fla. to Va. Linear-leaved aster (*A. linariifolius*), with numerous uniform, linear leaves and bright violet ray flowers is often associated with the large-flowered aster. Widely distributed, Fla. northward.

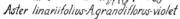

Aster linariifolius-A.grandiflorus-violet

pale violet or pinkish *violet-purple disk yellow*

Aster carolinianus A. adnatus

CAROLINA or CLIMBING-ASTER (*A. carolinianus*) A remarkable species of aster, a woody vine with widely spreading branches and trailing or climbing stems, finely and softly grayish-pubescent. Flower-heads solitary or clustered on the ends of copiously leafy branches. Ray flowers pale-purplish or pinkish, blooming all year. In swampy situations, Coastal Plain, Fla. to S.C. *Aster adnatus* is a remarkable and easily recognized aster because of its numerous, small, appressed and more or less overlapping branch leaves. The slightly spreading involucral bracts have dark-green tips. It frequents dry pinelands of the Coastal Plain from Fla. to Miss. and Ga.

DAISY-FLEABANES (*Erigeron*) The daisy-fleabanes resemble on the one-hand daisies and on the other Asters. They are annual or perennial spring bloomers with tall sparingly-branched stems. They differ from Asters in having the involucral bracts in only one series, and the heads terminate leafless stalks. Some are weedy, growing on road-shoulders, and in gardens and in fields. Heads have a yellow center and the rays are white, pink, or purple. Spring daisy-fleabane (*Erigeron vernus*) is a perennial with a rosette of thickish, club-shaped leaves and naked stems, divided above into a loose corymb of white to lilac heads. Acid soil in pinelands, Coastal Plain, Fla. to La. and Va. *E. quercifolius* is a common roadside weed with lobed basal leaves.

basal leaves vary-1-2

[136]

Erigeron vernus - pinkish to white

ROSIN-WEEDS, ROSIN-PLANTS (*Silphium compositum*) This tall perennial with a wand-like stem and large basal leaves with long petioles and deeply-lobed blades is often seen in summer on forest margins in various districts from Ga. to Ala., n. to Tenn. and s.e. Va. The almost leafless stem branches above into several long, spreading branches, terminated by relatively large flowers with large yellow rays. Rosin-plants are so called because many of them have a resinous juice. Only ray flowers develop seeds. Ga. to Ala., n. to Tenn. and s.e. Va.

Silphium compositum-yellow

GOLD-IN-GREEN, GOLDEN-STAR (*Chrysogonum virginianum*) *Chrysogonum* means "golden knee" in allusion to its golden-yellow flowers and knee-like joints of its runners. Since the "knees" are not golden, a more appropriate name is gold-in-green, suggested by Dr. B. W. Wells of N.C. State College. This low, early-blooming, perennial composite is a favorite with many lovers of wild flowers, and when grown in rock gardens and borders it is a striking ornamental. Dry, open woods, various localities, from Fla. to La., n. to Tenn. and Pa.

Chrysogonum virginianum - yellow

GREEN-EYES (*Berlandiera*) Berlandieras are erect perennials with deep roots and alternate leaves with toothed or lobed blades. Heads are more or less nodding, with broad, short bracts and yellow rays. As in rosin-weeds, only the outer or ray flowers perfect seed. *B. subacaulis* is distinct from all other species in the South in its hispid stem and deeply-lobed leaves. In the other 2 species the leaves are merely toothed or only lobed at base and gray-woolly hairy. *B. pumila* has a leafy stem while in *B. humilis* it is conspicuously leafless above. The first 2 occur only in pinelands of Fla. but the last ranges in the Coastal Plain from Fla. to Ala. and N.C.

[137]

yellow

Berlandiera subacaulis-B.humilis

WILD-QUININE (*Parthenium integrifolium*) A curious, herbaceous perennial, growing in dry, mostly clayey soil in open areas or open, often pine, woods. The basal leaves are long and coarse, gradually decreasing upward. Several small, grayish-white heads are arranged in a corymb at the end of the stem. The heads are unusual in that the customary 5 rays are very short. *P. argentatum,* a shrub native of Mex. and Tex. called guayule, is grown in the arid Southwest as a source of rubber. *P. Hysterophorus,* an annual adventive from tropical America called Santa Maria, occurs in pinelands, cultivated and waste ground, from Fla. to Tex., n. to Pa.

white

Melanthera hastata Parthenium integrifolium

Tetragonotheca helianthoides-yellow

PINELAND-GINSENG (*Tetragonotheca helianthoides*) A sunflower-like composite with large heads and yellow rays. Its generic name comes from the Greek words meaning "four-angled" (*tetragonos*) in allusion to its 4-angled involucre and "cup" (*thece*) referring to the cup-like base of the united involucral bracts. All flowers bear "seed" and each one is enclosed in a folded bract (chaff). Although it resembles a sunflower, it is more closely related to rosin-weeds (*Silphium*). It is a conspicuous plant of dry, sandy woods and pinelands, blooming nearly all summer. Coastal Plain, Fla. to Miss., n. to Tenn. and Va. The only one of its kind found in the s.e. U.S.

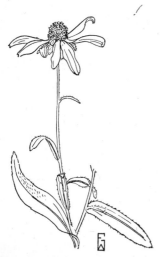

Rudbeckia divergens- orange×brown

CONE-FLOWERS, RUDBECKIAS (*Rudbeckia*) The Rudbeckias are called cone-flowers also and in many respects resemble *Echinacea*. However, the rays are always yellow, but often darker at base, and the chaff is not rigid. There is considerable variation in appearances between different groups of species. The best known is perhaps *R. hirta,* or black-eyed Susan, which is widely distributed in the eastern half of N.A. Other species, some of which are cultivated, closely resemble this one. *R. divergens* is a southeastern species characterized by being divergently branched, growing in pinelands and fields, Coastal Plain, Fla. to S.C. This is one of the species used by the Seminoles for fever and headache.

[138]

PHOEBANTHUS *(Phoebanthus tenuifolia)* *Phoebanthus tenuifolia* is one of the only 2 species of this genus found in the world. The other is *P. grandiflora* which occurs in peninsular Fla. while the former is found only in n. Fla. They are perennial, sunflower-like herbs, differing from sunflowers mainly in their flat achenes ("seeds"). Both have relatively large flower heads with both disk and ray flowers yellow. They may be distinguished from most sunflowers by their long, narrow leaves, although *P. tenuifolia* is quite similar in its leaves to the narrow-leaved sunflower *(Helianthus angustifolia)* which, however, has a dark center. They grow from horizontal tubers which are easily transplanted to the wild flower garden. Both grow in relatively dry situations.

Phoebanthus tenuifolia yellow

CROWN-BEARDS *(Verbesina laciniata)* Crown-beards have few (4-5), relatively large ray flowers which in this, as well as another species of wider distribution *(V. occidentalis)* are white. They are mostly perennial herbs but may be shrubby. Heads are numerous and being clustered are quite showy. The achenes have 2 short barbed awns which stick readily to cloth or hair. *V. laciniata* is a conspicuous flower during most of the summer in pinelands and hammocks, Coastal Plain, Fla. to S.C.

Verbesina laciniata - white

SEA-OX-EYE *(Borrichia frutescens)* This is a low, fleshy, shrubby composite which grows in salt marshlands on the coast with salt-marsh grasses and glass-worts. It is sparingly branched with opposite, entire, rounded leaves and the whole plant has a grayish appearance. The heads are relatively large but the yellow rays are short. In certain situations, it grows in large patches in almost pure stands. It blooms in late summer and fall. Fla. to Tex. and s.e. Va. A larger and more variable species *(B. arborescens)* grows in s. Fla. and the Keys; also in tropical America.

[139]

yellow

brown

Helianthus radula · Borrichia frutescens

Helianthus debilis - yellow

white
Bidens pilosa

yellow
B. mitis

white

yellow

Marshallia obovata Endorima uniflora

SUNFLOWERS (*Helianthus*) *Helianthus* ranks among the larger genera of the Compositae, such as Asters and goldenrods. Small lists 43 species in his *Manual of the Southeastern Flora*. They are usually tall, annual or perennial herbs, with simple, alternate or opposite leaves, native to the Americas. About 100 species are known. They are characterized by having large heads; in the cultivated common sunflower it may reach several inches across. The beach-sunflower (*H. debilis*) is a southern annual, frequenting mostly coastal sand dunes from Fla. to Tex. and Ga. *H. radula* is a remarkable species with a simple stem up to the inflorescence and the rays are so short they seem to be absent altogether. Sandy pinelands, Coastal Plain, Fla. to Ala. and Ga.

BUR-MARIGOLDS, BEGGAR-TICKS (*Bidens*) Among the assortment of fruits and seeds that stick to your clothing as you wander in the autumn woods you will no doubt find the fruits of *Bidens,* sometimes also called "Spanish-needles." Most species have yellow rays which in some weedy species are inconspicuous, but in the s.e. *B. mitis,* they are unusually large and showy. Another beautiful, tropical and subtropical species is the white-flowered *B. pilosa,* known commonly as "shepherd's-needle," usually found near the coast in sandy soil, blooming all year. Fla. to Ala. and s. Ga.; tropical America.

BARBARA'S-BUTTONS (*Marshallia*) As in the cultivated ragged-robin, the florets of Barbara's-buttons are all tubular, and Marshallias are sometimes called "wild ragged-robins." They are, however, quite similar in size and shape and are mostly white to pink or purplish. They are quite ornamental and when seen in great abundance present a beautiful sight. *M. obovata* is a spring-blooming species, mostly in the Coastal Plain but extending to the lower Piedmont. Fla. to Ala. and N.C. A variety of this extends to Va.

ENDORIMA (*Endorima uniflora*) This is a conspicuous, yellow composite of sandy soil in low pinelands of the Coastal Plain, characterized by having a relatively tall, unbranched stem terminated by a single, fairly large, yellow head. It is further recognized by the partly united chaff bractlets forming a cup-like structure surrounding each achene ("seed"). It ranges from Fla. to La. and N.C. Another species (*E. atropurpurea*) with a dark-purple center, occurs in s. Ga.

[140]

BITTER-WEED (*Helenium tenuifolium*) The commonest composite seen during the late summer and autumn as it has invaded road shoulders, pastures, and waste grounds everywhere. In small, overgrazed pastures and barnyards, it has become so dominant that they often show a solid yellow. Cattle avoid it, but if eaten by cows it turns the milk bitter, hence the common name. The name *Helenium* is supposed to have come from Helen of Troy of Greek mythology. Heleniums are generally called "sneeze-weeds." Actinospermum (*Actinospermum angustifolium*) is a monotypic genus, meaning it has only one species, a showy annual or biennial with yellow heads, blooming in fall or all year. It has top-shaped seeds margined above with several small, rounded scales. Fla., Miss. and Ga.

COREOPSIS, TICKSEEDS (*Coreopsis*) Several species of *Coreopsis* occur in the southern states, some seen more often than others because they occur in disturbed ground, such as roadsides, margins of fields, and waste places. Such are usually migrants from other lands. One of these is *C. grandiflora* which in recent years has rapidly migrated from the vicinity of coastal areas inland along roadsides so that in some sections the road-shoulders are yellow with them in late spring. This species and its varieties and close relatives are often cultivated in gardens. Another introduced species which is also escaping to similar situations is *C. tinctoria,* with dark-red disk flowers and yellow rays with crimson-brown base. All species of *Coreopsis* have involucres in 2 series, of which the outer one is green and spreading.

POLYPTERIS (*Polypteris integrifolia*) This is another monotypic genus inhabiting dry pinelands in Fla. and Ga. It has small, pink or white heads of tubular flowers only. A tall perennial with alternate or opposite, thickish leaves. Hymenopappus (*Hymenopappus carolinensis*) is a remarkable composite in that the bracts of the involucre resemble the flowers in color (petaloid), giving added showiness to the heads. The florets are all tubular and at first white or pink, changing promptly to a purplish-red. The plant is a biennial or perennial erect herb with much dissected leaves. It grows in dry, sandy soil of various provinces from Fla. to Tex., Kans. and S.C.

[141]

yellow   yellow
Helenium Tenuifolium Actinospermum angustifolium

center red-brown   yellow
Coreopsis tinctoria   C. grandiflora

white or pinkish   disk flowers purplish
bracts white
Polypteris integrifolia Hymenopappus carolinensis

yellow
Arnica acaulis   Gaillardia lanceolata

BLANKET-FLOWERS, GAILLARDIAS (*Gaillardia*) Gaillardias are distinctive in their soft center from the long, slender chaff between the disk florets. The most often seen is G. *pulchella,* meaning "the little beauty," which originally ranged from Tex. to Fla. but has spread to waste places in all coastal cities as far n. as s.e. Va. It shows remarkable variation in color and shape of ray flowers. Arnica (*Arnica acaulis*) has been considered to have medicinal value since the herbalists of the Middle Ages. Only one species occurs in the South. It has a few relatively large yellow flowers at the summit of a stem, although the specific name means "without a stem," and a rosette of leaves at base. It is fairly frequent and locally abundant, although Small says, "Rather rare." Acid soil in open woods, Coastal Plain, Fla. to Pa.

red
milia coccinea Senecio vulgaris

GROUNDSELS, RAGWORTS, SQUAW-WEEDS (*Senecio*) These spring-blooming perennials have, with few exceptions, small, yellow heads in flat-topped inflorescences, and alternate, variously lobed leaves. The most common native species is S. *Smallii* which in recent years is spreading along highways. A still showier species is an introduced annual (S. *vulgaris*) which grows in great profusion in moist, open ground. Widely distributed in e. N.A. and on the Pacific Coast. Red-emilia (*Emilia coccinea*) is a native of the Old World tropics which has invaded cultivated or otherwise disturbed ground in southern Fla. and the Keys. The tubular flowers in a vase-like involucre look like little red brushes.

purple   mauve
Cirsium horridulum   C. Nuttallii

THISTLES (*Cirsium*) Thistles are better known for their vicious spines than for their beautiful pink, violet, lilac, purple, or yellow brush-like heads. The individual flowers have long, slender tubular corollas, crowded in heads with spiny involucres. One of the most conspicuous is yellow-thistle (C. *horridulum*), well-named for it is horribly spiny. A variety, frequently seen in the southern states, has purple flowers and was named var. *Elliottii* for the S.C. botanist, Stephen Elliott. Various situations, Tex. to Fla., n. to Me. C. *Nuttallii,* a more typically Coastal Plain species, bears the name of one of the early botanists of this country, Thomas Nuttall (1786-1859). It has a mauve-colored head and the leaves are white-woolly beneath. Sandy soil, Fla. to La. and s.e. Va.

[142]

SUN-BONNETS (*Chaptalia tomentosa*) This is one of the earliest flowers to bloom in the Coastal Plain, associated with white-violets, pitcher-plants, and sun-dews. The plant is grayish-hairy and the flowers of the nodding heads are purplish outside and white within. As in stokesia, the flowers vary from completely strap-shaped marginally to tubular in the center. Tex. to Fla. and N.C. A species with larger heads (*C. dentata*) occurs in the Everglade Keys. Flowering-straw, "rose-bush" (*Lygodesmia aphylla*) belongs to the dandelion type of composites, of which there are several interesting and beautiful examples in the South. The most often seen is the blue-flowered chicory (*Cichorium Intybus*) which is naturalized along roadsides and in waste places. The flowers open in the forenoons and remain closed the rest of the day. Flowering-straw is somewhat similar with reduced leaves but with rose-colored flowers, rarely white. Dry pinelands, Fla. to Ga.

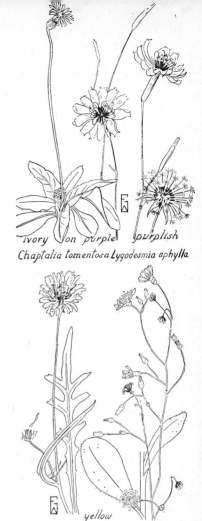

ivory on purple   purplish
Chaptalia tomentosa Lygodesmia aphylla

FALSE-DANDELION (*Pyrrhopappus carolinianus*) This resembles a tall dandelion but with leafy stems, most often seen along roadsides and in waste places. Also like the dandelion its achemes grow a long, slender "beak" with the silky sail at the summit. A variety with cream-colored corollas, tipped with rose, occurs locally in n. Fla. and Ala. Several species of hawk-weeds (*Hieracium*), ligulate-flowered composites occurring in the South, bloom at different times of the growing season. Most of our species have yellow flowers. Perhaps the best known is the spring-blooming poor-Robin's plantain (*H. venosum*) with a basal rosette of purplish-green leaves with light-green veins. *H. Gronovii* is a late summer to autumn bloomer, named for a Dutch botanist, Jan Fredrik Gronovius (1690-1762).

yellow
Pyrrhopappus carolinianus-Hieracium Gronovii

SOW-THISTLE (*Sonchus oleraceus*) An introduced annual, growing as a weed in cultivated grounds, with numerous yellow heads of flowers. Although not especially ornamental, its odd shape with its spiny, clasping leaves often arouses curiosity. Widely distributed in e. N.A. Grass-leaved lettuce (*Lactuca graminifolia*) is represented by several species in the South. Tall annuals or perennials, they have mostly yellow flowers, but one (*L. hirsuta*) has purplish flowers and in the one illustrated they are a shade of purple-blue. The latter is found in fields and open woods, Fla. to Tex. and N.C.

[143]

yellow   purple-blue
Sonchus oleraceus Lactuca graminitolia

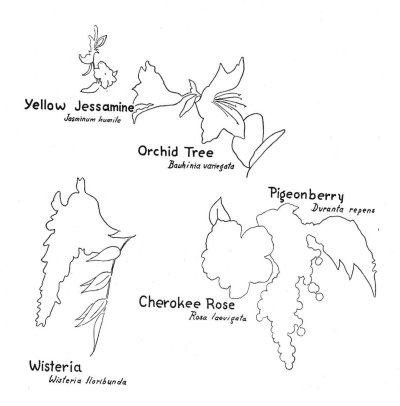

Yellow Jessamine
Jasminum humile

Orchid Tree
Bauhinia variegata

Pigeonberry
Duranta repens

Cherokee Rose
Rosa laevigata

Wisteria
Wisteria floribunda

SUBTROPICAL SPRING ORNAMENTALS
FROM COAST TO COAST

SUBTROPICAL SPRING ORNAMENTALS FROM COAST TO COAST
Bauhinia, yellow jessamine, Cherokee rose, Duranta (flower and fruit), wisteria

NATIVE ORCHIDS AND AIR PLANTS ON A TREE
IN THE FLORIDA EVERGLADES

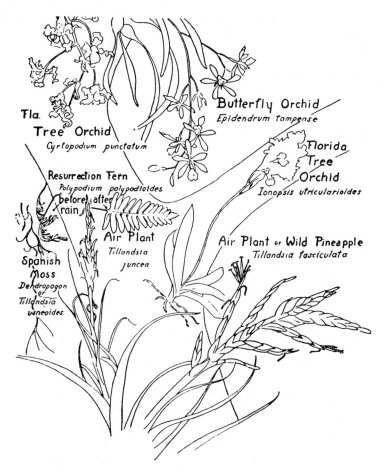

NATIVE ORCHIDS AND AIR PLANTS

All of these plants are classed as epiphytes from their habit of living on trees but deriving no nourishment from them. They are, therefore, not parasites as the mistletoe, and do not cause any apparent injury to the trees upon which they are lodged.

Although some of these species are native of different sections and may not be found growing on the same tree, they have been brought together outside in s. Fla. in the manner indicated in the composite painting opposite. Likewise, while all do not usually bloom at the same time, it is quite possible for them to flower together as out of season blooming is so prevalent in Fla.

The sale of Tillandsias as orchids has caused great confusion as to their names. In spite of the law for conservation of native plants, both Tillandsias and orchids are being sold and bought by people who do not know how to grow them, with the result that they are fast vanishing. Orchids take so many years to grow from seed and are so particular in their requirements as to humidity and warmth that they should not be handled or sold by the inexperienced unless he is in a position to give them the care they need.

Tillandsias can be easily distinguished from orchids by their stiff grey-green scurfy foliage, crowded in rosettes at the base, resembling the cultivated pineapple, while orchids have delicate graceful racemes of irregular flowers and leathery dark green leaves rising from clumps or pseudo-bulbs.

In the competition for space, nutrients, air, and light, plants have adapted themselves in various ways. One of the most unique of such adaptations enables the plants to habitually grow on other plants without being in contact with the soil and deriving no water or nutrients from the plants upon which they grow. These curious plants are known as epiphytes, meaning "upon plants." They have sometimes been mistaken for "parasites" and the plants upon which they grow, for "hosts." Since water is one of the most critical substances for plants and is usually obtained from the soil, the greatest structural specializations of epiphytes are necessarily those related to the absorption and conservation of water obtained from other sources than the soil. That these structural adaptations are, however, limited is indicated by the fact that the epiphytes which represent the more highly differentiated plants occur mostly in the tropical rain forests where rain and atmospheric moisture are most abundant. Here the trunks, branches, and even leaves of woody plants are densely covered with epiphytes ranging from the microscopic algae and fungi to lichens, mosses, a wide variety of ferns, aroids, bromeliads, and orchids. Some of these may not spend all of their lives as epiphytes, such as the ancestor of the Boston fern (*Nephrolepis exaltata*) which starts on the ground but soon sends out rapidly growing runners which sooner or later ascend trees on which new plants are formed and become flourishing epiphytes. This represents a "nest epiphyte" which develops an extensive entanglement of roots by means of which it accumulates vegetable debris for nutrients and which acts as a sponge for water absorption. Another type of epiphyte is represented by the southern grey polypody fern which curls up when dry, thus reducing loss of moisture, but when wet unfolds to a normal shape. For this reason it is sometimes called the "resurrection fern." Relatively few families of flowering plants are represented among epiphytes, but a few of these consist mainly of epiphytes. One such family is the Bromeliaceae or pineapple family. Most of the bromeliads are of the "tank" type of epiphyte in that the spoon-shaped bases of the closely overlapping leaves store considerable quantities of water. The roots are mainly for holding on to the support, be it a tree, a post, or a wire. Here belong the noted "air plants" of Florida which are exceptional epiphytes in not being adapted to a tropical rain forest. The most remarkable of all bromeliads is Spanish-moss which has no roots whatsoever and simply hangs over branches in long festoons. It absorbs and conserves water by means of scale-hairs which give it its grayish appearance. Just how it obtains enough nutrients for its subsistence is still somewhat of a mystery. Epiphytic orchids have usually two kinds of roots, some are specialized for absorption of water and the others for anchorage. The excess water absorbed is stored in bulb-like structures. Since most of the commercially grown orchids are epiphytes they are not grown in soil but in plant fiber which often consists of entanglements of fern roots known as osmundine. So far as known, the only composite epiphyte is *Senecio parasiticus* of Mexico, which, as the name indicates, was once thought to be a parasite.

Exotic

While many wild flowers and ornamental plants do not have what the flower arranger calls "distinction" suitable for show arrangements, many, when selected to harmonize with an indoor color scheme, are most useful and attractive for decoration in the house when other flowers are scarce. They, however, must be picked with care and imagination. If one is careful in planning foundation and border planting in the South, one may have a succession of flowers for the house throughout the year. Mixed bouquets of exotics on pages facing 65, 144, and 177 show many flowers that are effective but rarely used for cutting. (Most bloom in warm sections from coast to coast.) None of these attempt to be "arrangements" but were combined to illustrate as many as possible in color and to suggest ones that can be easily used for decoration. Most are overcrowded. The wild flower composite bouquets and groups on pages facing 17, 32, 48, 56, 57, 64, and 129 show many of the wild flowers that are the most worth picking and that last the best. A few orchids and lilies are too rare to pick in some localities, while in some places they are so abundant that a few might be picked.

CUTTING HINTS: To prevent wilting the following suggestions are given to help the amateur enjoy more flowers in the house. They apply equally to exotics and wild flowers.

Pick early in the morning or late in the afternoon.

Put immediately in deep water. (Carry a can in the car for wild flowers.)

Condition the most fugacious by floating in a dishpan of water one or two hours before arranging or float vines in low bowls. (Flame vine, thunbergia, and pandora.)

Scrape the stems of woody plants and seal the stems of those with milky juices by immediate immersion in boiling water or by waxing. (See poinsettia, page 112.)

Select sprays having the most interesting lines and do not overcrowd (as has been the tendency in these composite bouquets); three sprays carefully chosen are often more effective than great bunches. A single specimen is often interesting.

Conserve our wild flowers by picking only a few and never cut the rare ones.

Use scissors to avoid disturbing the roots; prune wild shrubs to shape them, leaving buds and enough flowers to re-seed themselves. Everyone should be informed as to which flowers are on the conservation list in his state and should refrain from picking them. A few which are easily grown may be transplanted. Because of the roadside bulldozer, conservation ideas have changed recently in some localities. For instance, in Louisiana, when a new road is being laid out, garden club women are encouraged to bring Irises and spider lilies to wet garden spots so they may be preserved.

To select unusual material is a challenge. Berries from ornamentals are available most of the year. Dried material can be kept on hand for use at anytime or shipped north. This includes seed pods of

Brazilian pepper, koelreuteria, legumes, wood roses, eucalyptus, bottle brush, and wild flowers and grasses; foliage of monkey-puzzle, sea grape, and magnolia. (Magnolia leaves and those of many other ornamentals may be soaked weeks in a solution of 1 part glycerine and 2 parts water to give a lustrous brown.) Cultivated flowers include congea, assonia, bryophyllum, and strelitzias.

ARRANGING: Strip off excess leaves before arranging as most ornamental and wild flowers have too many leaves and so look weedy. Select a vase suitable to the material and color scheme. (Containers of bamboo, copper, brass, and rustic materials blend well.) Use a good flower holder to keep the stems in place or use chicken wire crumpled, strips of lead, sand, or moss. Place the flowers carefully in the container one at a time or in small clumps, when masses of color are desired. Avoid overcrowding and remember the principles of balance. Simplicity and informality are the keynote for wild flowers. Some plants that have a particularly artistic line of growth can be strikingly arranged after the Japanese manner. Three or five sprays make an interesting group. Grasses may add variety. When more than one color is used, mass the colors to avoid a spotted effect. Small ornamentals and wild flowers or parts of large panicles are good for miniature arrangements. These provide good practice for children, cultivating their originality and sense of proportion.

PLANTING: A wild flower garden is a joy. Most of the varieties in the composite bouquets may be transplanted to the wild garden. Orchids, Catesby's lily, and lupines may be transplanted only if given a *similar environment* to that in which they grow naturally. Generally annuals are more easily grown from seed, while perennials should be transplanted. Other small varieties which can be successfully cultivated are: iris, lobelia, lantana, sand verbena, obedient plant, *Croton alabamensis*, buckwheats, crotalaria, blue sage and red sage, basil, halesia, erythrina, milkweed, passion vine, wild poinsettia, andromeda, pitcher plants, climbing aster, Carolina jessamine, most lilies, Rhododendrons, Azaleas, sea grape, and heaths.

Many of our wild flowering trees and shrubs are cultivated in the North where they are highly prized. Garden clubs should encourage their use for ornamental and roadside planting in the South, but most should be purchased from nurserymen as these plants have developed root systems that withstand transplanting. Thanks to garden clubs, dogwood and red bud decorate the southern streets. Other suggestions are flowering plums, crabs and haws, Hydrangeas, Magnolias, gordonia, franklinia, sour wood and sumac, fringe tree, Rhododendrons, Azaleas, and heaths.

ARUM FAMILY (Araceae)

CERIMAN (*Monstera deliciosa*) This tropical American aroid is unparalleled as a decorative house plant for its bold, dark, deeply cut, and perforated leaves. A spectacular vine in the far South, it climbs vigorously to the tree tops. The creamy white spathe resembling that of the calla-lily, folds around a thick spadix densely covered with greenish bisexual flowers. This spadix becomes the fruit and is delicious to eat when the hexagonally designed coat peels off. Zones 1 and 2. Recovers from frost damage. Propagated from cuttings; thrives best in rich, moist, half-shaded, protected locations.

Monstera deliciosa - white

AGAVE FAMILY (Agavaceae)

CENTURY PLANT (*Agave americana*) Closely related to the lily and amaryllis families, these striking big plants, with a rosette of huge, gray-green, thick succulent leaves, give a tropical effect to formal modern landscaping. They mature in 20 to 30 years, and after developing a tall stem 20'-40' high, they blossom and die. Young plants develop in the inflorescence and take root when the old plant becomes heavy and falls, while other species reproduce by suckers at the base. Some forms of *Agave* are used for the heavy hemp-like fibers in their huge leaves, and acres of them are planted in Mexico for *pulque,* the intoxicating, native drink made from its sap. Zones 1 and 2.

Agave americana- yellow-green

YAM FAMILY (Dioscoreaceae)

AIR-POTATO (*Dioscorea bulbifera*) Beware of the air-potato! It was introduced into this country by the Division of Plant Exploration and Introduction, U.S.D.A., hoping that its many bitter, aerial tubers could be used medicinally. It was not only found worthless but the vine has become a nuisance because of its aggressive growth and rapid propagation by the aerial tubers. Many are attracted by the beauty of its heart-shaped leaves, but it should be carefully watched when cultivated in warm sections. Originally introduced as S*milax sandvicensis* and later, from Hawaii, as *Dioscorea divaricata*.

Dioscorea bulbifera -yellowish

AMARYLLIS FAMILY (Amaryllidaceae)

RED SPIDER-LILY (*Lycoris radiata*) The dainty red *Lycoris* is the outstanding fall-blooming bulb throughout the middle South. The spidery blossoms with long, protruding stamens come before the linear leaves and make a gorgeous show in clumps or borders. This introduction from Japan and China is used extensively in Zones 3-4 but may be grown in Zones 1-2. They last well when picked.

Lycoris radiata - red

GIANT AMARYLLIS (*Amaryllis vittata*) This famous amaryllis is found in gardens throughout the lower South in great numbers, especially in Ala., Miss., and La. It has huge, dark red, trumpet-shaped flowers with a fine white stripe part way down the petals. The modern hybrid amaryllis are distant stepchildren of *A. Johnsonii*, which reputedly was the first hybrid amaryllis, created in England by a Lancashire watchmaker named Johnson around 1800. The Mead strain, an orange-red variety which originated in Fla., is more commonly grown only on the peninsula. Bulbs of choice imported Dutch varieties of many shades are grown in pots for flower shows and in special beds, but they are too expensive and tender for indiscriminate outdoor planting. Rare types which are grown by specialists include the lavender-blue Worsleya Rayneri (*A. procera*) from Brazil and *A. Belladonna* var. *Haywardii*, the beautiful rose-pink novelty from Bolivia.

Amaryllis Johnsonii - deep red

AMAZON-, MADONNA-, or EUCHARIST-LILY (*Eucharis grandiflora*) The Greek word, *eucharis*, meaning graceful, aptly describes this purest of waxy-white flowers from the Andes of Colombia. A 5-pointed star, 3"-4" across, with a jonquil-like center, has a delightful fragrance. The blossoms are generally borne in 2's and bloom in winter and spring, although they may be forced to bloom at other times. A popular, bulbous houseplant, it thrives best in pots outside when rootbound. Zones 1-2. Plant in pots in shade; in sun for winter blooming. Unfavorably affected by division for propagation even in spring. Excellent focal point in arrangement.

[152]

Eucharis grandiflora - white

BANANA FAMILY (Musaceae)

BIRD OF PARADISE FLOWERS (*Strelitzia*) These stately members of the Musa or banana family, having odd, bird-like blossoms peeking out from boat-like bracts or spathes, are unrivaled for exotic motifs, with their long, wind-frayed leaves. *Strelitzia Reginae*, true bird-of-paradise flower, is well known for its brilliant orange and bluish flowers atop a pointed, purple-green spathe. A fall and winter bloomer, it grows in 3' clumps with blue-green leaves having reddish-midribs a foot long. Zones 1-2. Widely used by florists and in shows, it can be forced to bloom at will. *Strelitzia Nicolai* (color page facing 160) could be the "bird-of-paradise tree" for it develops a trunk. It should not be confused with *Ravenala*, for its leaves are less fan-shaped and young plants spring in clumps from rhizomes by means of which it is propagated. Its larger white and blue blossoms, from fall through spring, resemble a regal bird soaring out of a purplish spathe. The whole spathe is distinctive, but picking it spoils blooms to come. Individual flowers are unique and long lasting, even out of water.

Strelitzia Reginæ - orange x blue

Strelitzia Nicolai - white x blue

TRAVELER'S-TREE (*Ravenala madagascariensis*) This plant should not be called "traveler's-palm," for this Musa is not a palm. The name comes from the fact that water is caught in the leaf sheathes and gushes out when basal leaves are cut. It has a single trunk from which leaves radiate in flat, fan-shaped patterns. The brilliant peacock-blue seeds make spectacular arrangements and are so vividly colored they were thought to be artificial when judged at a flower show. Zone 1; protected parts of 2. Tender, needs rich soil, water and sun. Bank for frost protection. Appropriate for conservatory and patio.

[153]

Ravenala madagascariensis

Heliconia species

WILD-PLANTAIN, LOBSTER-CLAW (*Heliconia*) One of the most brilliant and bizarre of the big-leaved banana family, with 30 to 40 species native in the American tropics, but being tender summer bloomers are not too well-known even in Fla. The highly colored, boat-shaped bracts of glowing scarlet to yellow are geometrically arranged in fascinating designs. The largest species opposite is called lobster-claw because of the color and shape of its large bracts. The long-stalked leaves of 2'-3' rise on slender stems often from 8'-10' high and have decorative use for protected patios. Zone 1 and parts of 2. Long lasting for arrangement.

Hedychium coronarium-white

GINGER FAMILY (Zingiberaceae)

BUTTERFLY- or GINGER-LILY (*Hedychium coronarium*) Throughout late summer and fall, this tropical ginger from Asia gives us a succession of pure white flowers for decoration when other flowers are scarce. Several large, fragrant, butterfly-like flowers, 3"-4" across, stand erect from a terminal bract, while the old flowers droop. They are borne on tall spikes 4'-6' tall, with smooth, ornamental, canna-like leaves. Zones 1 and 2. Foolproof for lake shore. When potted, requires considerable water. Especially suited for church decoration and large effects. Thin leaves and remove old flowers.

Grevillea robusta - yellow; orange

PROTEA FAMILY (Proteaceae)

AUSTRALIAN SILK-OAK, SILVER-OAK (*Grevillea robusta*) The silvery-green foliage of this stately tree from Queensland is so beautiful and lace-like that small seedlings are used as decorative pot plants. Prized in Australia for cabinet wood, we believe it has possibilities here commercially in the subtropics. A quick-growing hardwood, it attains 100' and a diameter of 2½' in a little over 20 years. Dwarf species of *Grevillea* are extensively used in Calif. The vivid orange and yellow flowers with an elongate pistil are borne in elongate, one-sided racemes. Zones 1 and 2. Will endure 15°. Easily seeded, but because of long tap roots is difficult to transplant.

BIRTHWORT FAMILY (Aristolochiaceae)

PELICAN-FLOWER, BIRTHWORT (*Aristolochia grandiflora*) The humorist of plants, its bud caricatures a pelican, swan, or goose, while the gaping mouth and enlarged calyx of velvety mahogany entice the carrion fly for pollination by emitting a vile, nauseating odor; but it is not insectivorous. Whimsically enough, it is used medicinally to strengthen the stomach and appetite! A gigantic Dutchman's-pipe (18″-2′ across), from Guatemala, with huge seed "baskets" and a long hairy tail, up to 3′, from the calyces. By contrast, the graceful vine which bears these flowers hangs in spirals and has lovely, heart-shaped leaves softened beneath with silky hairs. Zone 1. From seed or as an annual from cutting. Needs support.

Aristolochia grandiflora—yellow, maroon

BUCKWHEAT or KNOTWEED FAMILY (Polygonaceae)

CORAL VINE, ROSA DE MONTANA, CORALLITA (*Antigonon leptopus*) (color page facing 65) One of the world's most beautiful climbers, coral vine is at home over the lower South into Calif. Its clusters of tiny rose-pink flowers are borne in huge festoons over fences or shrubbery during the fall and winter months until cut down by frost. Lettuce-green foliage repeats the shape of the flowers on a larger scale. The flower-sprays spring intermittently from the leaf axils. Requires little care; withstands 10°. Propagates from seed, cuttings, tuberous roots. Good for greenhouse or conservatory. Effective hanging from tall vase or mantel or in flat bowls.

Antigonon leptopus - pink

POKEWEED FAMILY (Phytolaccaceae)

AGDESTIS (*Agdestis clematidea*) In fall, this graceful vine is covered with dense axillary panicles of dainty, sweet-scented, white to greenish, 4-sepaled blossoms with many stamens. Petals are lacking. Because of the vile odor of its attractive, wide, heart-shaped, grey-green leaves, it should be planted some distance from the house. Fine for covering unsightly objects, it is a quick-growing, branching vine from C.A., attaining 50′ in one season. Zones 1 and 2. From a rock-like or turnip-shaped tuber (to 100 lbs.). Rich soil and moisture; propagated by offsets. Lovely and long lasting in arrangement when stripped of foliage.

[155]

Agdestis clematidea- white

Bougainvillea *hort. var.* Afterglow

Nandina domestica - white

Magnolia Soulangeana - white, pink

FOUR-O'CLOCK FAMILY (Nyctaginaceae)

BOUGAINVILLEA, PAPER-FLOWER (*Bougainvillea glabra*) (color page facing 97) The glorious bougainvillea is perhaps the most popular woody vine for all warm climates. The magenta-purple *B. glabra* and its many hybrids are useful for varied purposes, responding well to shears. The spectacular red *B. spectabilis* var. *crimson lake* is a more robust grower. Recent hybrids can be had in almost any shade and hue from white through the yellows, oranges, purples, pinks, to deepest red. Profuse and almost perpetual bloomers, the brilliant clusters of 3-parted, heart-shaped, paper-thin bracts overshadow the tiny yellow tubular flowers in their center. Zones 1 and 2; protected parts of 3. Will endure 20°. Any soil; full sun.

BARBERRY FAMILY (Berberidaceae)

HEAVENLY-BAMBOO (*Nandina domestica*) For decorative effects on borders of lawns or foundation planting this slender dwarf shrub is a favorite in the mid-South and Calif. as well as in its native Japan. In early summer the reed-like stalks are topped by erect panicles of small 6-petaled flowers, followed by brilliant clusters of scarlet berries which last all winter. The graceful evergreen foliage of compound leaves gives a lacy appearance to the slow-growing clumps from 2'-10'. Tinged with bronze in spring and fall, the foliage turns deep-coppery red in winter. Zones 1-6; sun or shade (will survive 0°). Several plants are needed to produce berries.

MAGNOLIA FAMILY (Magnoliaceae)

CHINESE- or SAUCER-MAGNOLIA (*Magnolia Soulangeana*) Pride of gardens the world over, in early spring this glorious deciduous small tree of 10'-20' from China is covered with magnificent large blossoms of rosy-pink to reddish-purple petals, creamy-white within. Blossoms appear before the large bright green leaves but often along with the leaves on older plants. The star-magnolia, with star-shaped blossoms 3"-4" across, is also a popular and hardy ornamental. Hardy. Zones 3-5. Many of our native southern magnolias deserve cultivation. (See color page facing 1 and page 37). Distinctive for line arrangements in Oriental manner.

[156]

SWEET-SHRUB FAMILY (Calycanthaceae)

ORIENTAL SWEET-SHRUB (*Chimonanthus praecox*)
Closely related to our native sweet-shrub (*Calycanthus floridus*), this attractive introduction
from China and Japan blooms as early as
Jan., and earlier, hence its Greek name meaning
"winter-flower." Its bare twigs are dotted with
dainty pairs of bright yellow, sweet-scented,
"double" blossoms $\frac{3}{4}$"-1$\frac{1}{2}$" across, resembling
those of our native sweet-shrub. Their inner
sepals are striped with the same purplish brown.
A loosely-branched, scraggly shrub with large,
lustrous, opposite oblong leaves 3"-5" long.
Zones 3-6. Hardy to Philadelphia. Propagated
by seed, layering, or division. Thrives in well-
drained, fairly rich soil.

Chimonanthus praecox - yellow

ROSE FAMILY (Rosaceae)

FLOWERING-, JAPANESE-QUINCE (*Chaenomeles lage-
naria*) From Cen. Asia, this is a popular shrub
throughout the South except in subtropical regions. In late winter and spring, its brilliant
display of waxy-white, pink, orange-to-scarlet
pear-like blossoms come before the leaves. Rarely
over 5', its glossy, semi-evergreen foliage and
somewhat thorny habit of growth make it an attractive shrub. *Chaenomeles* is the name now
used, *Cydonia* being reserved for the fruiting
quince. The many hybrids have led to hopeless
confusion of names. Zones 3 and 4. Hardy
where peaches grow. From seed or root cutting.
Appropriate for Oriental arrangement.

Chaenomeles lagenaria-white, red

CHINESE-PHOTINIA, HOLLY-LEAF (*Photinia serrulata*) Most Photinias, stunning large woody
shrubs from China (20'-30'), are evergreen in the
South where they are prized for their handsome
foliage and for the variety they give to ornamental planting. In the spring the young foliage
is tinged with bronze; the large corymbs or clus-
ters of rosaceous blossoms $\frac{1}{4}$"-$\frac{1}{2}$" are white; the
glossy foliage of rather large leaves is a rich green
through the summer, turning deep red in fall;
the showy clusters of bright scarlet berries in late
fall are beautiful at Christmas time. All Photinias are useful for screens and heavy mass
planting. Zones 1-6, best for 3-6. Propagated by
seed, cutting, or grafting. Prefers light sandy
loam and sun.

[157]

Photinia serrulata - white-pome red

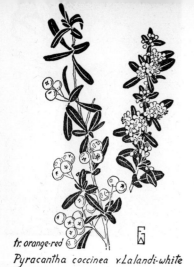

fr. orange-red

Pyracantha coccinea v.Lalandi-white

FIRE-THORN, PYRACANTHA (*Pyracantha crenato-serrata*) Fire-thorn is aptly named, for it is covered with long spines and fiery-red or orange berries through fall and winter. Berries set well in Fla. only in cold winters. This hardy, woody evergreen shrub from 10'-12' of spreading habit with small dark green leaves rounded at the tip, often serrate or crenate, comes from Europe and Asia. In early spring it is covered with clusters of white rose-like blossoms resembling *Cotoneaster*. *P. coccinea* with bright red berries and small leaves is perhaps the most popular throughout the South. Zones 1-5, best in 3-5. Almost any sunny location. Prune severely before transplanting.

yellow

Crotalaria retusa C.spectabilis yellow

PEA or PULSE FAMILY (Leguminosae)

RATTLEBOXES (*Crotalaria spectabilis* and *C. retusa*) (color page facing 177) Showy annuals from tropical America with luxuriant racemes of bright, glossy, canary-yellow flowers larger than the sweet pea. They are little appreciated in Fla. as they were used as a cover crop in orange groves to bring nitrogen to the soil and often escape from cultivation. The long, oval, grey-green leaves are single. Zones 1-3; an annual northward. Easily grown and self-seeding. Useful for picking in the fall and attractive in arrangement when excess leaves are removed. Occasionally sold by florists.

Bauhinia variegata - purplish or white

ORCHID-TREE, MOUNTAIN-EBONY (*Bauhinia variegata*) The butterfly tree is a superb sight for six weeks of late winter and early spring when in full bloom. Large orchid-colored butterflies with wings spread for flight are apparently on every twig of this grey-barked small tree (which is often nude of foliage at blossoming time). One of the five mauve to lavender petals is traced with cerise and based with white; hence the misleading common name, orchid tree. *B. candida* var. *alba* adds allurement to the white garden. The 2-lobed leaves generally follow the blossoms, and later long, reddish-brown seed pods develop. Many new species have been introduced. Zones 1 and 2. Grow from seed. Prune after blooming. Pick in cool of the day, crush stems, condition in deep water. Buds open inside.

[158]

WOMAN'S-TONGUE TREE, SIRIS TREE (*Albizia Lebbeck*) This small tree, from Asia and Africa, is known as "woman's-tongue tree" because it is covered with long, thin, tannish seed pods from late spring throughout the winter which rattle in the breeze. Prized as a shade tree for the lawn, it is quick growing with an attractive, spreading shape and graceful feather foliage, giving an effect similar to that of the less hardy poinciana. In spring, its fluffy pompons of greenish-yellow stamens resemble those of the pink *Albizia Julibrissin*. Zones 1 and 2. Quick growing from seed in almost any soil. Seed pods effective for dried bouquets.

Albizia Lebbeck-cream pompons

BIRD-OF-PARADISE SHRUB (*Poinciana Gilliesii*) The brilliant crimson stamens protruding 4"-5" beyond the clear-yellow petals give the effect of a bird poised on the tip of the branch. It should not be confused with *Strelitzia Reginae* which is called bird-of-paradise flower. Loose racemes, or clusters, of these flamboyant blossoms are set off against a background of feathery, extremely fine leaflets. Its delicate form, with scraggly branches and subtle texture, gives this shrub or small tree the ornamental effect of a dwarfed Oriental tree. Zones 1 and 2. Stands 20° with little damage but loss of leaves. From seed soaked in warm water; tolerates poor soil and drought. Keeps well when cut.

Poinciana Gilliesii-yellow-st. red

RINGWORM-CASSIA (*Cassia alata*) Huge poker-like spikes of brilliant canary-yellow flowers in the spring make *Cassia alata* a most ornamental shrub. It is an excellent filler 6'-8' tall. Though individual flowers do not open wide, they disclose the characteristic sickle-shaped pistil of Cassias, which later becomes a long-winged, scalloped-edged seed pod. Decorative, frond-like compound leaves are made up of 12-28 oblong leaflets over 3" long. Medicinally, one variety of ringworm-cassia is used to check vomiting and diarrhea. Zone 1, parts of 2. From seed in arid regions or well-drained sunny spots. Withstands salt spray.

[159]

Cassia alata - yellow

ANGEL'S TRUMPET (*Datura arborea*)   Nightshade Family

This aristocratic relative of the common Jimson-weed, from Peru and Chile, is periodically covered with pendulous, pearly white flowers 8"-9" long, resembling trumpets. They open first at night and have a musk-like odor. Both the leaves and seeds are poisonous when eaten in large amounts. Zones 1 and 2; parts of 3, from cutting. In N.C., store roots in cellar over winter. Grows in any soil; blooms even in shade. Prune after each flowering. Subject to root-knot. Effective decoration if floated in water to prevent wilting.

EASTER LILY-VINE (*Beaumontia grandiflora*)   Dogbane Family

This vine is a breath-taking sight when in full bloom during the spring, for its fragrant clusters of milky-white blossoms are as large, as beautiful, and more graceful than the Easter-lily it resembles. It is often tipped with pink and veined with green. This woody climber from India with heavy foliage of long, supple, leathery leaves needs support, but given room it "takes over" if not pruned severely. Zones 1 and 2. Though tender, when once established, it is not badly hurt by frost in protected locations in c. Fla. Propagated by layering. A flower show winner when selected judiciously, conditioned, and pruned of excess leaves.

BIRD OF PARADISE TREE (*Strelitzia Nicolai*)   Banana Family

This is a most unusual plant for creating modern effects in ornamental planting as well as flower arrangements. It may develop a trunk and eventually become tree-like up to 25' in height. Its long, blue-green banana-like leaves, 3' or 4' long and 1' wide, are no less striking and decorative than its long white blossom with a blue tongue, which resembles a white bird soaring out of a large purplish, boat-like spathe. Only recently have these unique, long-lasting flowers been recognized as appropriate for arrangements. Zones 1 and 2.

CHALICE-VINE, CUP OF GOLD (*Solandra guttata*)   Nightshade Family

From Mex. and the W.I. comes this quick-growing woody climber. Exquisitely fashioned like a goblet, flowers may grow to 10". Buds, protected by a green calyx at the base, are greenish white, creamy to yellow as they open and turn to gold before they fall; they are delicately traced with purple. At night they spread a fresh cocoanut fragrance. Large oval leaves, shiny and leathery, form the foliage. Aerial roots require strong support; the vine rarely climbs more than 30'. Zone 1, parts of 2. Tender. Best in sandy loam. Start young shoots in sand. Fugacious unless flower touches water or has water inside.

Angel's Trumpet

Easter Lily-Vine

White Bird of Paradise

Chalice-Vine or Cup of Gold

Combretum

Chorisia

Sprouting-Leaves

Air Plants or Bromeliads on Palmetto

COMBRETUM (*Combretum coccineum*) Combretum Family

In late winter and spring, this showy, woody, free-flowering vine from Madagascar, with large oblong leaves, is covered with glorious terminal, one-sided clusters of brilliant vermilion, tubular flowers with protruding stamens. Its leathery fruits are winged and decorative. Zones 1 and possibly 2. Propagated by cutting firm wood. Lasts well in arrangement.

CHORISIA (*Chorisia crispiflora*) Bombax Family

This large-flowered small tree, native of the American tropics, is seldom seen in Fla. and Calif. but is worthy of more extensive cultivation, for it is hardy in c. Fla. and stands abuse. Its long buds suddenly burst open disclosing a mallow-like pistil and 5 beautiful rose-colored petals, white at the base mottled with rose and yellow. It is closely related to the yellowish flowered floss-silk tree (*C. speciosa*) whose silky floss on the seeds is used for stuffing pillows.

SPROUTING-LEAVES (*Kalanchoë*) Orpine Family

K. verticillata is a fascinating succulent for it is a leaf-propagating biennial, reproducing clusters of tiny plants at the tips of its long tubular leaves. Whorls of thick leaves, mottled with purple, make an interesting geometric design. In mid-winter, long-lasting clusters of brilliant coral tubular blossoms dangle from its top. *K. Fedtschenkoi,* a graceful, succulent "scrambler," 1'-2', with opposite, iridescent, grey-green, obovate leaves with dentate margins, is striking planted at its base for contrast. It is similar but smaller; blossoms of light-salmon flower as the other wanes. Easily reproduced. Stands abuse, poor soil, sun and drought. Zones 1 and 2 or in pots. Withstands light frosts.

AIR PLANTS or BROMELIADS Pineapple Family

Most of these bromeliads are recent introductions from Brazil by Mulford Foster of Orlando, Fla., and all are hardy winter and spring bloomers for c. Fla. and Calif. (*mid-right*) *Billbergia pyramidalis* has a glorious, though short-lived, brilliant spike of pink to scarlet bracts with blue-tinged petals. (*upper left*) *Aechmea* hybrid is "Foster's favorite," the first bromeliad to be patented. A berry-like orange flower stalk with midnight blue blossoms droops from clusters of light-green leaves seemingly lacquered with wine red. (*lower left*) In April, *Neoregelia* sp. "crimson cup" blossoms. Its rosette of bright red leaves lasts several months but the flowers are less conspicuous. (*center*) *Vriesia carinata* hybrid Marie, "painted feather," has a most spectacular flat flower spike of brilliant red and yellow that blooms from September for six months. All are easily grown inside in medium sandy soil. They require little watering as they hold water at the leaf base.

[161]

Daubentonia punicea - scarlet

SESBAN, RED-LOCUST, PURPLE-SESBAN (*Daubentonia punicea*) Drooping flower clusters of brilliant orange-scarlet (rarely rose-purple) make this large, branching, leguminous shrub a striking color note for the back of borders as it blooms freely most of the warm weather, but its compound leaves are deciduous in some areas. Its resemblance to the locust gives one common name, but purple does not seem descriptive of its color. Introduced to Fla. from S.A. it has spread rapidly, becoming naturalized along roadsides from Tex. to Va. Zones 1-3. Easily grown from seed. Likes sun and any soil.

Thryallis glauca - yellow

MALPIGHIA FAMILY (Malpighiaceae)

THRYALLIS (*Thryallis glauca*) This neat, low shrub from tropical America, is popular for planting in front of foundations because of its low bushy growth and continual, profuse bloom. While the upright clusters of small (¾″), thin, 5-petaled, deep-yellow flowers with red stamens are not conspicuous, they lend color to any planting. The fine, wiry stems of glossy, red-brown add contrast to the small, blue-green foliage. Many showier shrubby vines of this family with pink- or yellow-clawed and ruffled flowers, though tender, would prove successful for Zone 2, as several species are native in Zone 1 in Fla. and Tex. Zones 1 and 2; deciduous in 3. Easy in any soil and drought resistant. From cutting or seed. Lasts well for small bouquets.

Stigmaphyllon littorale - yellow

BUTTERFLY-VINE (*Stigmaphyllon littorale*) Daintiest of slender-stemmed, medium-sized, woody vines of the American tropics, though tender, they should be used more in warm climates. In spring axillary umbel-like clusters of golden-yellow, sprite-like blossoms of 5 round, clawed petals with ruffled edges, cover these charming vines like a cloud of tiny butterflies. The crisp, heart-shaped, ivy-like foliage of the most used *S. ciliatum* is ciliated or hairy and makes an excellent trellis vine. The species illustrated has thrived in c. Fla. for over six years. Attractive and long lasting when cut.

[162]

SPURGE FAMILY (Euphorbiaceae)

COPPER-LEAF, BEEFSTEAK-ACALYPHA (*Acalypha Wilkesiana*) These shrubs, 12'-15' in height which came from the S. Seas, are probably the highest-colored and most popular ornamental foliage plants for warm climates. Miles of them are seen as hedges in Puerto Rico and Hawaii. The curling, serrated, almost heart-shaped leaves, mottled with red, purple, or white give an excellent color note. Many varieties with mosaic markings have been developed. Inconspicuous, tufted, red flowering spikes droop from the leaf axils, but in the chenille plant (*A. hispida*) they are often 2' long, hanging in brilliant red chenille-like curls. Zones 1 and 2; also greenhouse. Easily grown from cutting. Full sun. Although tender, they recover quickly from frost. Prune often.

Acalypha Wilkesiana-red leaves

TUNG-OIL TREE (*Aleurites Fordii*) These decorative small shade trees are generally seen in n. Fla. in groves where they are cultivated extensively for tung-oil, an important drying oil for fine paints, varnishes, and lacquer, which is extracted from the thick-shelled nuts. Introduced by the U.S. Dept. of Agriculture from China and E.I., it thrives in n. Fla. and s. Ga. in soil too poor for farming. Spring tips the ends of the nude branches with large panicles of 5-petaled, many-stamened, white flowers streaked with red, just as the long petioled, palmately-veined, smooth-edged, ovate leaves appear. Parts of Zone 3. Easily grown from seed in light frost areas.

Aleurites Fordii   reddish-white

CASTOR-OIL PLANT, CASTOR-BEAN (*Ricinus communis*) One of the oldest plants under cultivation, it is easily grown and decorative, despite the fact that castor oil is made from its seed! Its staminate and pistillate flowers with rough seed pods are unusual, and huge palmately-lobed red or green leaves with toothed edges are beautiful. A native of Afr., it grows as an annual up to 15' tall and branches like a tree. Zones 1-3; an annual in 4-5. Quickly grown from seed (planted inside in the North). Rich, well-drained, sandy loam, full sun. Thrives in poor soil in Fla. For arrangement, condition young growth in water; lasts well.

Ricinus communis (monecious)

Koelreuteria bipinnata - yellow-pod pink

SOAPBERRY FAMILY (Sapindaceae)

GOLDEN RAIN-TREE (*Koelreuteria bipinnata*) Even though deciduous, the more hardy *K. paniculata* is extensively used in Zones 3 and 4 for ornamental planting because of its handsome, compound foliage, yellow flowers, and clusters of inflated, three-valved, pointed seed pods. Both medium-sized trees, native of China and Japan, are topped with iridescent pink to orange-red as the seed pods age. Their tiny, yellow, four-petaled flowers, borne in spring and summer, are similar. The shattering of their petals accounts for the common name, golden rain-tree. Zones 1-3; parts of 4. Self-seeding *K. bipinnata* is less hardy. Papery seed pods decorative and long lasting for dried bouquets.

Tetrastigma Harmandii-brown berry

GRAPE FAMILY (Vitaceae)

TETRASTIGMA (*Tetrastigma Harmandii*) One of the most decorative of the woody grapevines, its handsome thick leathery foliage of 3-5 leaflets powdered with brown are arranged finger-fashion. Though the flower panicles are inconspicuous, the long-lasting clusters of edible, russet-brown, round grapes, similar to the scuppernong, are decorative. Introduced from the Philippines by the U.S. Dept. of Agriculture but not used commercially. Zone 1; parts of 2. Tender, grown in protected spots from seed, cutting, or layering. Appropriate for patios and greenhouses.

STERCULIA FAMILY (Sterculiaceae)

PTEROSPERMUM (*Pterospermum acerifolium*) A close relative of the chocolate, this lovely tree from tropical Asia has proved hardy in most of Zone 2. In late winter to early spring the long, banana-shaped, rust-covered buds unfold, disclosing a large, pure white, tubular flower with 5 rolled petals 5"-6" across. The unusually wide oak-like leaves, often 14" x 12", are silvery grey-green and slightly tomentose and brown underneath. The Greek name refers to its winged seeds. Zone 1 and protected parts of 2. Although nipped by 20°, it is an ornamental tree well worth growing throughout the citrus area. Needs warmth and perfect drainage. Propagate by cuttings of half-ripe side shoots.

[164]

Pterospermum acerifolium-white

TEA FAMILY (Theaceae)

SASANQUA-CAMELLIA (*Camellia Sasanqua*) Though smaller-flowered and hardier than *C. japonica*, no flower surpasses the beauty of the single sasanqua form, which blooms so profusely so early (Sept. through Dec.) and so young, or bears such thick, glossy foliage. There are more than 100, often fragrant, varieties introduced from Japan, from single to double, from white to purplish-red, from dwarf to vigorous growers. Var. *Cleopatra* is covered with dainty, pale-pink, semi-double blossoms. Zone 3; parts of 2 and 4. Roots easily from cuttings; thrives in sandy, moderately acid soil; likes afternoon sun and liquid manure. Transplant in winter. May be espaliered or trimmed as hedge. Pick carefully to shape bushes.

Camellia Sasanqua var Cleopatra-pink

TEA-SHRUB (*Thea sinensis*) Closely related to the camellia, this commercial member of the tea family has pretty, white, axillary blossoms like tiny, single camellias, with conspicuous yellow stamens and a pleasing fragrance. Its leaves produce the tea of commerce, for which it is grown in subtropical areas and highlands throughout the tropics. An evergreen shrub or small tree, to 30', its attractive flowers, dark evergreen foliage, and bushy, dense growth make tea popular for ornamental planting as well, and it is deserving of more extensive cultivation in this country. Withstands 15°. Thrives through the camellia belt and has the same culture.

Thea sinensis-white or pink

LOOSESTRIFE FAMILY (Lythraceae)

CRAPE-MYRTLE (*Lagerstroemia indica*) Grown in mild climates the world over, this shrub is widely cultivated in the South for roadside beautification. A dependable, profuse bloomer throughout summer and fall, it is often called "The Lilac of the South." The plants with smooth stems are pleasing even when leafless. Trained as a small dooryard tree or a flowering shrub, it blooms best when pruned after flowering. Varieties bear either white, pink, lavender, or purple blossoms; the most popular is watermelon-red. Zones 1-3; parts of 4. Easy to grow from seed or cuttings. Effective when cut even though the flowers shed. Buds open indoors.

[165]

Lagerstroemia indica-white -pink-red

Hylocereus undatus - white

CACTUS FAMILY (Cactaceae)

NIGHT-BLOOMING CEREUS (*Hylocereus undatus*)
Truly the grail among flowers, few are as beautiful, as spectacular, or as fairy-like when their
glistening white petals suddenly unfold on a still
moonlit night, disclosing a cross-like pistil surrounded by numerous delicate yellow stamens.
Native of the W.I., this plant has become naturalized on palmetto trees along the Indian River
where it is breath-taking to see hundreds open at
one time. Uncanny, snake-like, triangular stems
make cereus decorative when not in bloom and
useful in creating modern tropical effects. The
red fruit, though a "prickly-pear" covered with
a greyish fuzz, is edible. A great variety of cereus
cacti are grown in warm climates and are both
suitable and interesting horticulturally for house
culture. The dainty *Epiphyllum* is especially
fine for c. Fla. for, during eight months of warm
weather, it is periodically covered with blossoms
lasting until the sun is high in the morning.
After putting them in the refrigerator, it is ex-
citing to watch them open in the daytime. Zones
1 and 2 or in pots. Practically epiphytic; effective in "bootjack" of palmetto.

Epiphyllum sp. - white

POMEGRANATE FAMILY (Punicaceae)

POMEGRANATE (*Punica Granatum*) Pomegranates
are among the showiest deciduous shrubs of the
middle South. They are covered with a profusion of brilliant scarlet blossoms from early May
through most of the summer and are equally
beautiful in the fall when their lustrous foliage
turns brilliant sunshine-yellow. A large well-
rounded shrub from Asia, there are many horticultural single and double, white to red varieties.
The decorative yellow fruit, the size of a small
orange, has innumerable seeds set in an acid
pink pulp and is used for jellies and drinks.
Zones 1-5. Easily grown from cutting; requires
sun but little or no care. Prune in late winter
for blossoming on current year's growth. Long
lasting in arrangements.

[166]

Punica Granatum - orange-red

MYRTLE FAMILY (Myrtaceae)

DOWNY-MYRTLE, HILL-GOOSEBERRY (*Rhodomyrtus tomentosa*) Beneath the under surface of gray-to-olive green leaves is a brownish down which gives the name downy myrtle to *Rhodomyrtus*. A symmetrical shrub of 3'-10', it is abundantly covered in late spring and summer with beautiful rosy pink flowers. The blue-black fruit, like huckleberries, is useful for pies and with guavas for a fine jelly. Originally from tropical Asia, Australia, and the Philippines, it is a favorite from Fla. along the Gulf to Cal. Zones 1 and 2. Withstands several degrees of frost. Sow seed in flats. Grows well in any soil with some irrigation. Difficult to transplant but naturalizes in moist locations through seed spread by birds.

Rhodomyrtus tomentosa - rose

BOTTLE-BRUSH (*Callistemon rigidus*) An unusual tall exotic shrub from Australia, it seems to bristle with deep-red torches from March to May. The gaudy spikes with conspicuous stamens give a spectacular appearance as these are more conspicuous than the tiny petals. The hardened close-set fruits resemble a cylinder of grey buttons below the developing branch beyond. A plant's age may be estimated by counting these fruiting portions. Several varieties with dense, willowy foliage are popular for lawns from Fla. to Calif. Zones 1 and 2. Drought resisting, rugged, dislikes transplanting. Loose, sandy soil. Long lasting when picked.

Callistemon rigidus - red

COMBRETUM FAMILY (Combretaceae)

RANGOON-CREEPER (*Quisqualis indica*) The blossoms of this rampant shrub-vine open white in the morning, strangely turn to deep crimson by twilight. Umbel-like clusters resemble stars falling from a bursting rocket, and these flowers seem to dangle in mid-air, but are attached by long, thin calyx tubes (3"-4") of yellow-green. A native of India, blooming from late spring into fall, its flowers sway gracefully from their support against long, dark-green, not too dense, opposite ovate leaves. Zones 1 and 2; from 5-angled seed or soft wood cutting. Prune severely after flowering. Pest free; protect from heavy frost. Long lasting for arrangement.

[167]

Quisqualis indica - white + pink + red

Tibouchina semidecandra-purple

MELASTOMA FAMILY (Melastomaceae)

GLORY-BUSH, PRINCESS-TREE (*Tibouchina semidecandra*) Glorious flowers, in terminal clusters, with 5-6 royal-purple petals, are conspicuous, not only for their size (3″-5″) but for their velvety texture and splendid, long, yellow stamens. A shrub or small tree of scraggly growth (to 10′), it will bush if pruned. Its pointed leaves, deeply veined underneath, of a soft texture, covered with white hairs, turn bronze-red before falling. A popular greenhouse plant, it often blooms for 8 months. Zones 1 and 2. Full sun. By cutting in spring; protect from frost, tender but recovers after 20° has apparently killed it. Shatters too quickly for picking.

ARALIA or GINSENG FAMILY (Araliaceae)

RICE-PAPER PLANT (*Tetrapanax papyriferus*) A most popular non-branching shrub from Formosa, this plant creates bold tropical effects for patio, foundation planting, or border. Huge palmately-lobed, long-stalked leaves over a foot across, tomentose with down beneath, grow palm-fashion from the top of a slender pithy stalk 8′-12′. In winter 2′-3′ branching panicles like graceful plumes blossom from the top, with round ball-like pompons of creamy white, 4-petaled flowerlets. Orientals make "rice paper" from the pith-like stem, which is used for drawing or, when dyed, for artificial flowers. Zones 1-4. Any soil, likes shade; difficult to transplant, but a nuisance in that it spreads too rapidly by off-shoots used for propagation.

Tetrapanax papyriferus-white

PLUMBAGO or LEADWORT FAMILY
(Plumbaginaceae)

BLUE-PLUMBAGO, CAPE-LEADWORT (*Plumbago capensis*) Unsurpassed for continuous bloom, this herb-like shrub came from the Medit. region and S. Afr. Dainty clusters of delicate, azure-blue, phlox-like flowers with longer tubes bloom in profusion when planted in the sun. Its small, light-green foliage gives pleasing contrast to foundation planting from Ga. to Calif. It blossoms twice during winter, as a house plant or in the greenhouse. Zones 1-2; part of 3. Drought resistant; well-drained, sandy soil. Effective in clumps, pruned as a hedge or trained to vine (20′). Propagate by division, cuttings of mature wood, seed. Pick carefully and condition to prevent wilting.

[168]

Plumbago capensis- azure-blue

OLIVE FAMILY (Oleaceae)

Jasminums, or jessamines, are the South's most popular woody vines and shrubs because of their profuse blooming and fragrant, star-shaped white or yellow flowers. Most are hardy, with protection, to Washington, D.C., and are easily pruned to shapely shrubs and hedges.

WHITE JASMINES, though more tender, are evergreen in the far South and Southwest and bloom almost continually. The slender catalonian (*I. officinale* var. *grandiflorum*), long grown in Europe for perfumery, is hardy but its blossoms hold on untidily. Nehrling's introduction, *J. illicifolium*, is popular in Fla. for its dark, evergreen foliage. Grand Duke (*J. Sambac*), a colonial favorite because of its fragrant, double flowers 2½" across, is hardy and a perpetual, though not a profuse, bloomer. Star jessamine (*J. gracillimum*), a rank grower and continual bloomer useful for screens, is easily grown, but its foliage often turns yellow.

YELLOW JASMINES are favorites in the deep South. *J. Mesnyi* is one of the commonest in the middle South. *J. humile* resembles it but requires support. *J. nudiflorum*, nude in winter, is covered with bright yellow blossoms at the first breath of spring. Few common names have led to more confusion, for many call any shrub or vine bearing fragrant, white, star-shaped flowers a "jasmine."

NON-JASMINES. No plant having a specific name *jasminoides* is a jasmine; this merely means jasmine-like. A few non-jasmines are also illustrated here so their flowers may be compared with the hope that they will no longer be confused with jasmines or called by that name. (They are described elsewhere.) Others often confused with jasmines are: tail-grape (*Artabotrys odoratissimus*), not cinnamon-jasmine; potato vine (*Solanum jasminoides*); *Murraya paniculata*, not orange-jasmine; *Cestrum Parqui*, not willow-leaved jasmine; *Mandevilla laxa*, not Chilean or star jasmine; Androsaces, not rock-jasmines.

[169]

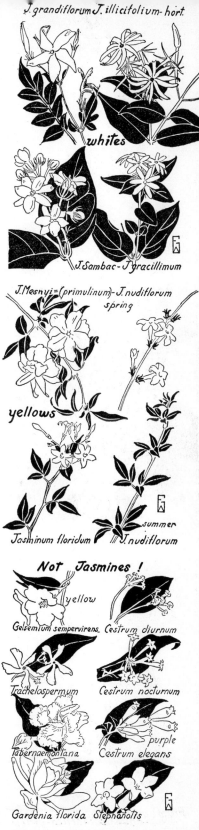

J. grandiflorum J. illicifolium- hort.

whites

J. Sambac - J. gracillimum

J. Mesnyi = (primulinum) - J. nudiflorum
spring

yellows

Jasminum floridum   J. nudiflorum
summer

Not Jasmines !

Gelsemium sempervirens   yellow   Cestrum diurnum

Trachelospermum   Cestrum nocturnum

Tabernaemontana   Cestrum elegans   purple

Gardenia florida   Stephanotis

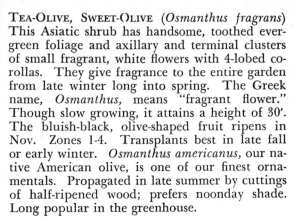

TEA-OLIVE, SWEET-OLIVE (*Osmanthus fragrans*) This Asiatic shrub has handsome, toothed evergreen foliage and axillary and terminal clusters of small fragrant, white flowers with 4-lobed corollas. They give fragrance to the entire garden from late winter long into spring. The Greek name, *Osmanthus,* means "fragrant flower." Though slow growing, it attains a height of 30'. The bluish-black, olive-shaped fruit ripens in Nov. Zones 1-4. Transplants best in late fall or early winter. *Osmanthus americanus,* our native American olive, is one of our finest ornamentals. Propagated in late summer by cuttings of half-ripened wood; prefers noonday shade. Long popular in the greenhouse.

Osmanthus fragrans - white

CALIFORNIA PRIVET (*Ligustrum ovalifolium*) Several partially-evergreen Ligustrums are the most popular broad-leaved shrubs for ornamental and hedge planting from coast to coast. They are used for their handsome glossy, leathery foliage of darkest green and many are pruned so often that their attractive blossoms are rarely seen. All have terminal panicles of small 4-petaled white flowers in late spring and summer, followed by a black drupe. *L. ovalifolium* to 15' is used extensively as a hedge. Evergreen *L. indicum,* with pubescent branchlets, is easily shaped and kept to a compact shrub for base planting and is one of the most widely used species. Evergreen in Zones 1-5. Foliage lasts well for decoration.

Ligustrum ovalifolium - white

DOGBANE FAMILY (Apocynaceae)

ALLAMANDA, GOLDEN-TRUMPET or -BALL (*Allamanda cathartica*) This vigorous climber, with brilliant yellow clusters of trumpet-shaped flowers, is a quick cover for walls and fences or may be trained as a medium-sized specimen shrub. Blooming profusely off and on except in very cold weather, the 5-lobed corollas of the striking var. *Hendersonii* often spread to 5" across. The handsome ovate, yellow-green leaves, 5"-6" long, have a glossy, leathery texture. Easily grown in Zones 1 and 2; or in the conservatory. Recovers quickly from freezes. Propagated from cuttings, layering, or division. Pick in cool of day, float in water to prevent wilting. Buds open in house.

[170]

Allamanda cathartica - yellow

NATAL-PLUM (*Carissa grandiflora*) This introduction from Natal is one of the most popular broad leaf evergreen shrubs for foundation planting in the subtropics for its glossy foliage, white star-like flowers, and ornamental red fruit. Its loose, spiny structure is decorative and it is easily shaped into a hedge or dense shrub. Sprinkled with fragrant, narrow-petaled, waxy-white flowers, 2″ across, it has the appearance of being covered with stars. The edible, decorative yellow to scarlet fruits, like inverted ruby drops, 1″-2″ long, make good jelly. Zones 1 and 2. Easily grown from cutting or layering; excellent for sea coast. Keep pruned. Likes moisture and warmth, not particular to soil. Foliage lasts well for decoration.

berry scarlet

Carissa grandiflora- white

OLEANDER (*Nerium Oleander*) The immemorial oleander which has shaded man's path since the Miocene age is now used in pots in the North and for boulevard plantings and as screens in the South. Of many varieties, huge clusters of single or double flowers of a vanilla fragrance range from the hardier white through pink to deepest red. Since early French days, New Orleans has been noted for its oleanders. Their milky juice is poisonous but sometimes used for snakebite and skin disease. A coarse, evergreen shrub, with long, narrow leaves, it reaches a height of 20′. Zone 1; parts of 3. Roots endure 10°. Mulch or store in cellar, Zone 4. Cuttings root easily. Yearly pruning and chlordane spray for caterpillars.

Nerium Oleander white, pink, red

CONFEDERATE-JASMINE (*Trachelospermum jasminoides*) This evergreen vine from Malaya with dark, glossy-green, leathery foliage has been the stand-by of old southern gardens since colonial days. Its perfume is even more fragrant than the jasmines with which it is frequently confused and it is incorrectly called "star jasmine," "Malayan jasmine," or "African jasmine." In early spring it is covered with panicles of white 5-petaled, pinwheel-like blossoms. Zones 1-3; parts of 4. Withstands 15°. Difficult to start from cutting; hardy. One of easiest trained and dependable vines for a trellis, tub, tall trees, low banks, or conservatory. Wilts quickly when picked.

[171]

Trachelospermum jasminoides- white

Ervatamia coronaria-white

FLEUR D'AMOUR, EAST INDIAN ROSEBAY (*Ervatamia coronaria*) This white-flowered shrub, 3'-5' high is often confused with jasmines because it is called "crepe-jasmine." Florists recently created its popularity for corsages by calling it "fleur d'amour." In spring and summer its clusters of beautiful double white flowers with crapey textured petals resemble gardenias though smaller and less scented, and the glossy leaves are lighter green and less leathery. It has the characteristic milky juice and pinwheel-like petals of the dogbane family. Excellent for base planting in frost-free areas. Zones 1 and 2. Quickly and easily grown from cutting. Prune frequently. Attractive to pick but condition in water or burn stem (to prevent wilting).

Hoya carnosa-pinkish

MILKWEED FAMILY (Asclepiadaceae)

WAX-PLANT (*Hoya carnosa*) This is one of nearly a hundred species of climbing shrubs from China and Australia which twine by means of aerial roots to as much as 8' high. The leaves are opposite, thick and fleshy, ovate-oblong, 2"-4" long; the flowers, in sessile or pedunculate axillary cymes, are white with a pink center and are about ½" across. The calyx is small, glandular at base within; the corolla, fleshy, deeply 5-lobed, the lobes spreading or reflexed. Zones 1 and 2.

Cryptostegia grandiflora-magenta

RUBBER-VINE (*Cryptostegia grandiflora*) Rubber-vine is cultivated in India for its milky juice that yields an inferior rubber latex. Its twisted flower bud resembles a morning glory, but the clear magenta flowers, 2" to 3" across, are 5-lobed, and they bloom profusely from spring to fall. Grotesque 3-angled pods enclose brown seeds with feathery, silky sails. Pruning makes this African woody vine with glossy dark leaves an effective shrub, but when planted in the shade, it will climb indefinitely. Zones 1 and 2. Frost-free areas, sea coast. From seed or cutting. Buds open indoors, last well. Pods for dried creations.

[172]

MADAGASCAR JASMINE, WAX-FLOWER (*Stephanotis floribunda*) Aristocrat of the milkweed family, borne in characteristic round umbels from the axils of the thick, leathery, emerald leaves, the 5-lobed waxen-white corollas with lobes turned back look like tiny stars on slender tubes 2″ long. Both flowers and leaves are so perfectly moulded they look artificial. This exquisite, twining, woody vine from Madagascar has suddenly become popular for bridal bouquets. Free-flowering in spring, or forced to bloom at will, they are long lasting and so delicately fragrant they have been distilled for perfume. (This fragrance wrongly wins for it the name "jasmine.") Zone 1; spots of 2. Tender; needs support and some shade. Slowly from cutting.

Stephanotis floribunda—white

MORNING-GLORY FAMILY (Convolvulaceae)

WOOD-ROSE, YELLOW MORNING-GLORY (*Ipomoea tuberosa*) When dried, the seed-pod and calyx resemble an exquisite wood carving. Its 5-parted calyx turns back, curving petal-like around the wooden capsule that protects the seeds. Lovely wood-rose is a rampant twining vine with large, yellow flowers and palmately divided leaves. A native of the tropics, it has become naturalized in Fla. and Tex. Zone 1; part of 2. From seed. Needs support. Outstanding for dried arrangements.

Ipomoea tuberosa - yellow

CHRISTMAS-VINE, SNOW-IN-THE-JUNGLE, BRIDAL-BOUQUET (*Porana paniculata*) Few vines combine such glorious flowers and graceful growth habit as this fall- to winter-flowering, woody climber from Africa, the Orient, and Australia. Each branchlet terminates in profuse panicles of delicately fragrant, tiny bell-shaped flowers, minute replicas of morning-glories. Its soft grey-green leaves, 3″-6″ long, are heart-shaped and hairy beneath. It often reaches a height of 30′. Zones 1 and 2. Though slow to start, a quick grower when established. Propagated by layering. Suitable for sea coast and dry soil. Lasts well in arrangement.

[173]

Porana paniculata - white

calyx white

Clerodendrum Thomsoniae - red

BAG-FLOWER, BLEEDING-HEART, (*Clerodendrum Thomsoniae*) This unusual and ornamental vine from W. Afr. has clusters of white, heart-shaped calyces which are more conspicuous and longer lasting than the fragrant, crimson corollas hanging like a drop of blood from their tips. Especially popular from Fla. to Calif., for they, like the northern bleeding heart, bloom intermittently when other flowers are out of season; flower in part shade, and bloom more profusely after a period of drought. Its heart-shaped, glossy, deep-green leaves are very handsome. Extensively used in greenhouses and as a house plant. Put out in summer. Pruning helps blossoming. Slow growing until established. Excellent for seashore. Seedlings have similar flowers but revert, the calyx being tinged red to magenta.

Clerodendrum paniculatum-scarlet

PAGODA-FLOWER, GLORY-BOWER (*Clerodendrum paniculatum*) Huge panicles of brilliant scarlet, long, tubular, 5-lobed blossoms make this one of the most spectacular of all Clerodendrums; it blooms from spring through fall. A native of Java, it is a soft, woody erect shrub of the verbena family, attaining from 5'-6'. Especially useful as a filler shrub, for it grows quickly and not too rampantly. Its large, angular, slightly cordate leaves, covered with soft hairs, are decorative in themselves.

GLORY-BOWER (*Clerodendrum fragrans*) Tight clusters of strongly-scented, double, pinkish flowers, or white-tinged crimson, look like old-fashioned nosegays. A semi-woody shrub, from 5'-8', with an over-abundance of large, coarsely tufted, pubescent, heart-shaped leaves, it is often confused with *Dombeya*. Zones 1 and 2. Dislikes temperature below 28°, but roots survive 20°. Grown from seed or cutting of half-ripened wood. Needs some sun to bloom but will take shade. Interesting for cutting when other flowers are scarce; lasts well.

[174]

Clerodendrum fragrans - pinkish

TUBE-FLOWER, TURK'S-TURBAN (*Clerodendrum indicum*) Turk's-turban is more conspicuous for its decorative clusters of red-purple to black berries than for its strange white, tubular flowers. Blooming in late spring to fall, the 5-lobed corollas nod from their slender corolla tube, 3"-4" long. The 2- to 3-parted blue-black berries are backed by a brilliant red, star-like calyx which persists all winter. They terminate the end of a tall spike which rises to a height of from 4'-6'. A non-branching, woody herb with large, dark leaves, having no preference for sun or shade. Zones 1 and 2; part of 3. So easily grown from seed it has escaped from cultivation in Fla. Seed clusters long lasting and unusual in arrangement with colored foliage.

Clerodendrum indicum- white

GOLDEN DEW-DROP, PIGEON-BERRY (*Duranta repens*) Striking is the contrast of the dainty terminal panicles of lavender, 5-petaled flowers with the large, drooping clusters of waxy, bright-yellow berries, for *Duranta* blooms off and on all year. The small, ovate leaves of this large shrub are on long spiny branches which droop gracefully with the weight of the berries. Mocking birds love its berries. Zones 1 and 2; parts of 3. From seed and cutting, roots withstand 20°; excellent for sunny corner of foundation planting. New growth should be continually tipped to harden wood so berries will not have to be pruned off in shaping plants. Lasts well in arrangement; flowers and berries effective together.

Duranta repens- lilac- fr. yellow

MANDARIN'S-HAT, CHINESE HAT-PLANT, (*Holmskioldia sanguinea*) (see color page facing 177) When trained as a shrubby vine, this plant is especially effective in fall and early winter, with graceful, cascading branches covered with clusters of odd, orange-red blossoms from the axils of the serrated oval leaves, resembling an inverted coolie's hat. The round membranous calyx is most conspicuous, while the intense red, tubular corolla protrudes from beneath like a trumpet. This supple, half-scandent shrub from Asia blooms in the fall or almost perpetually in Calif. and Fla., and should be used extensively for its gay colors and bizarre shape. Zones 1 and 2. Needs support, warm, sunny, protected area, and yearly pruning; water abundantly.

[175]

Holmskioldia sanguinea- red, orange

Reading from left to right—top row

George Frank—Large, bright-rose with deeper center; early.

Glory of Sunninghill—Large, orange-scarlet to vermilion; compact growth; very late.

Elegans—Medium sized, pale-pink; fast grower; early blooming.

Return to left

Formosa—Very large flowering, purplish pink to lavender; robust tall growing, with large luxuriant foliage; one of the easiest to grow but clashes with salmon and red shades; early to mid-season.

Maiden Blush—White deeply variegated with pink.

Mme. Van der Cruyssen—Brilliant rose; Belgium hybrid.

Fielders White—Large pure white; tall; good foliage; early blooming.

Return to left

Violacea rubra—Reddish lavender; tall growing; late flowering.

Iveryana—Large, white, striped pink; sometimes solid white or pink; late flowering.

Prince of Orange—Tangerine to fiery red; medium growth; mid-season to late.

Return to left

President Clay—Some classify as rose-pink, late blooming; others, orange to red-orange; early blooming; tall grower.

Pink-Pearl Kurume—Small pink flowers, hose-in-hose; compact dwarf-growing, resembles Snow-Kurume-white; hose-in-hose.

Duc de Rohan—Small to medium flower; light salmon to rosy salmon; early, free blooming for a long period; medium growth.

Return to left

Hinodegiri-Kurume—Small bright scarlet flowers; dwarf compact; late blooming.

Elegans—A smaller blossom than those on the upper right.

Appleblossom-Kurume—Small white, tinged with pink; dwarf compact growth; late.

Pride of Mobile—Similar to Elegans, but larger flowers; lavender pink; mid-season.

Jean Haerens and another double Belgian hybrid; large double, rosy red flowers.

Since the color and size of azalea blossoms vary with conditions of climate and sunlight, season of bloom, and age of the blossoms, it is difficult to reproduce their colors accurately. Most of the blossoms here reproduced are a shade lighter than in nature.

INDICA AND KURUME AZALEAS

SUMMER FLOWERING EXOTICS

Subtropics from coast to coast

Dwarf Poinciana
Caesalpinia or Poinciana pulcherrima

Cape Honeysuckle
Tecoma capensis

Yellow Elder
Stenolobium stans

Golden Trumpet
Allamanda cathartica v. hendersoni

Bottle Brush
Callistemon rigidus

Crotalaria
Crotalaria spectabilis

Yellow Caesalpinia
Poinciana pulcherrima

Double Yellow Hibiscus
a Reasoner hybrid

Flame of the Wood
Ixora coccinea

Mandarins Hat
Holmskioldia sanguinea

Single Hibiscus
variegated hybrid

Baby Bauhinia
Bauhinia racemosa or Calpini

Shrimp Bush
Beloperone guttata

Fountain Bush
Russelia juncea

Leaf of Yellow Elder

SUMMER FLOWERING EXOTICS

[177]

Congea tomentosa - pinkish

CONGEA (*Congea tomentosa*) Certain plants have their "vogue," but congea was so in demand when first introduced that wealthy women fought over pieces from flower shows! From Burma and Malay, it is a strong woody, climbing shrub, related to *Petrea*. Earliest spring clothes it with huge branching panicles of 3-parted, petal-like bracts, more showy than the minute, white flowers they surround and lasting for months. Soft, silken hairs subdue their mauve-pink color to an opalescent velvety texture. The coarse, evergreen heart-shaped opposite leaves, hairy as they age, are often in poor condition. It has turned spots of Jamaica into a veritable fairyland, with a haze of mauve, covering the tallest trees. Zone 1; parts of 2. Happy in c. Fla. for 6 or 8 years. Effective on palm trees.

Petrea volubilis - purple on blue

QUEENS-WREATH, PURPLE-WREATH (*Petrea volubilis*) Queens-wreath is one of the most unusual of the woody vines. Native of Cuba and other islands of the W.I., and Brazil, it seems to bear 2 kinds of blossoms at once. Drooping racemes 8"-10" long of showy blue-violet, star-shaped calyces long outlast the true verbena-shaped flowers of velvety royal-purple which sit upon them. Zone 1; parts of 2. Plant in protected sunny southern exposure; keep trimmed. Difficult to start from cutting or layering. Excellent for conservatory.

NIGHTSHADE FAMILY (Solanaceae)

PARADISE-FLOWER, GIANT POTATO-CREEPER (*Solanum Wendlandii*) In spring and through summer and fall, huge lilac-blue clusters of potato-like blooms hang from the ends of paradise-flower. Individual blossoms are often 2½" across, in large forking clusters. Large, smooth-edged deciduous leaves, prickled underneath, may be simple, lobed, trifoliate, or compound and are often 10" long. One of the showiest and most robust climbers of the nightshade family, this Costa Rican native twines upward or is held by occasional prickles. Zones 1 and 2. Tender. Easily grown in light moist soil. Give support, full sun. Prune in winter. Handy when a cluster of blue is needed for arrangement; effective floating.

[178]

Solanum Wendlandii-lilac-blue

NIGHT-BLOOMING CESTRUM (*Cestrum nocturnum*)
Few plants have been so often confused with jasmines, for it is habitually called "night-blooming jasmine." Its heavy night odor is almost unbearable if planted too near the house. A woody shrub, 10'-12' tall with yellow-green foliage, it is periodically covered from spring to fall with racemes of inconspicuous, greenish-white, tubular flowers, followed by decorative waxy-white berries.

white berries
Cestrum nocturnum *greenish*

DAY-BLOOMING CESTRUM (*Cestrum diurnum*)
This somewhat hardier shrub of similar growth has daytime clusters of white, tubular flowers with rounded corolla lobes followed by black berries. Because of its heavy fragrance it is also often confused with jasmine. Zones 1 and 2.

Cestrum diurnum *-white, berries black*

PURPLE-CESTRUM (*Cestrum purpureum*) A graceful, not too vigorous, shrub from Mexico, this cestrum has smooth, green leaves and intermittent, nodding clusters of rose-purple flowers over 1" long. In shape the individual blossom suggests an inverted urn. It is often of somewhat climbing habit. Zones 1 and 2. Suitable for the house or greenhouse farther north. It likes sun but tolerates shade and light soil. It may be grown from seed or cutting taken in February or March. Lasts well when cut.

[179]

Cestrum purpureum - *purplish red*

Brunfelsia americana-white

LADY-OF-THE-NIGHT (*Brunfelsia americana*) "Dama de noche," is indeed fitting, for often at full moon and periodically throughout warm weather, it is covered with long white tubular flowers having a heavy, pungent odor. The five-lobed corolla has a thin, graceful tube (3″ to 4″) and turns creamy-yellow with age. Flowers fall in graceful cascades and are effective and long-lasting for picking. An attractive shrub from 4′-8′ tall, with glossy dark, leathery foliage, native of Mex. and the W.I., it should be cultivated more. A fine conservatory subject where it is usually a winter bloomer. Zones 1 and 2; and greenhouse. Easily grown and propagated by cuttings.

Paulownia tomentosa - violet

FIGWORT FAMILY (Scrophulariaceae)

PRINCESS-TREE (*Paulownia tomentosa*) The large clusters of tubular, violet flowers of the princess-tree resemble a *Jacaranda,* though not so handsome, so blue, or so long-lasting. A hardy, deciduous tree introduced from the Orient, it is becoming naturalized in the middle South. Named for a princess of the Netherlands, it blossoms for a short time before the large, heart-shaped leaves appear. Throughout the winter, the ovoid seed pods persist on bare limbs, giving an untidy appearance. Zones 3 and 4. Quick growing, tall tree. Naturalizing in Ala. as it grows so easily from seed.

Russelia equisetiformis-red

FOUNTAIN-BUSH, CORAL-PLANT (*Russelia equiseti-formis*) (color page facing 177) A pleasing departure from the usual run of shrubs, this airy Mexican plant with its thin rush-like, pendulous, nude branches is covered most of the year with a shower of 1½″ firecracker-like, tubular flowers of burning red, hanging in branching cascades. The whorls of minute leaves are often scale-like and tinged bronze. A flowering fountain, it is suitable for window-boxes, conservatories, and rockeries; as a potted plant; and for planting around pools, foundation, or front of borders. Zones 1 and 2. Sandy, dry, sunny location. Easily grown from cuttings and division; self-layering. Excellent alone as a shower bouquet. Lasts well.

[180]

BIGNONIA FAMILY (Bignoniaceae)

PINK TRUMPET-VINE, PANDORA (*Podranea Ricasoliana*) This evergreen climber, native to S. Afr., was admired by the ancients and was named for Pandora, who had beauty and grace. Exquisite branching panicles of funnel-shaped pink flowers, 2″-3″ long, hang gracefully in huge open clusters from the ends of the serpentine-colored sprays; 3 of the 5 rounded petals are veined with magenta down into the tube. It flowers intermittently throughout the year. The lacy leaves of 7- to 11-toothed leaflets, arranged feather fashion, are especially beautiful. Zones 1 and 2. Recovers from 20°. Slow growing at first. Recovers from frost quickly. A heavy feeder; requires root room, rich soil, sun, and support.

Podranea Ricasoliana-pink-red lined

LOVE-CHARM (*Clytostoma callistegioides*) This lovely pink bignonia from Brazil and Argentina (formerly called *Bignonia speciosa*), is perhaps one of the hardiest introductions of the bignonia family, as it withstands some frost. Its pale-lavender blossoms with yellowish tubes streaked with purple have indeed a "beauteous mouth," as implied by its Greek generic name. The funnel-shaped flowers, 3″ long, are borne in 2's in March to May from the pairs of glossy, dark, evergreen leaves. Unlike the genus *Bignonia*, it has prickly pods and simple tendrils by which it climbs. Though a strong grower, it does not get out of hand easily and can be trained. Zones 1 and 2; parts of 3. Foliage often withstands 20°. Lasts well in arrangement.

Clytostoma callistegioides - lav.

YELLOW-BIGNONIA, YELLOW ELDER (*Stenolobium stans*) (color page facing 177) A popular, small, flowering tree for the border, most conspicuous in fall and late winter, but tipped at intervals with a profusion of showy panicles of 2″ clear yellow, 5-lobed trumpets of typical bignonia-shaped flowers. The yellow-green foliage, of 5 to 13 pinnately compound, serrate leaves gives a light feathery appearance and is its only similarity to elder, with which it should not be confused. Long, narrow seed pods are 6″-8″ long. Hardy in Zones 1 and 2; parts of 3. Recovers from 20° or less; easy culture from seed or cutting; thrives near large trees. Blooms in shade. Prune to increase bushiness.

[181]

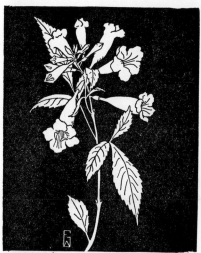

Stenolobium stans - yellow

Tecomaria capensis-scarlet

CAPE-HONEYSUCKLE (*Tecomaria capensis*) (color page facing 177) One of the handsomest half-climbing, shrubby bigonias (not a honeysuckle), it is prized for its luxuriant, glossy, evergreen foliage (of 7-8 compound, toothed leaves), as well as for its showy, upright, terminal spikes of vivid orange-red, trumpet-shaped, tubular corollas (2"), with conspicuous stamens resembling those of flame-vine but of a more brilliant scarlet. It blooms best from Aug. to Nov., and off and on at other times. Slow growing, easily pruned as a shrub, it clambers without tendrils to 25'. Graceful on a palm or wall, it can be trimmed as a hedge. Zones 1 and 2; protected parts of 3. Easily grown from cutting, layering, or seed in sandy soil; likes sun; responds to care.

Chilopsis linearis - pinkish-lav.

DESERT-WILLOW (*Chilopsis linearis*) This decorative small tree, 10'-20', is native to the arid regions of our Southwest from s. Tex. to Calif. and Mex., but is useful for street and other ornamental planting in the Southeast from Fla. through most of Zone 3. Its willow-like foliage is deciduous in some areas, but it is prized for its early summer-blooming clusters of bignonia-like light-lilac 5-lobed, crimped, trumpet-shaped blossoms, (1"-2"), yellow-striped within. The Greek generic name means "lip-like." Should be grown more; especially good for planting in parkways or in poor, dry soil. Grown from seed from the $\frac{1}{4}$" thick pod 6"-12" long.

ACANTHUS FAMILY (Acanthaceae)

BEARSBREECH, WHITE-ACANTHUS (*Acanthus montanus*) The acanthus leaf is famed for giving us the design for Corinthian pillars. White- to purplish-flowered *Acanthus montanus* is a medium-sized herbaceous perennial with long, spiny, leathery leaves, deeply pinnately cut and mottled. While the tall spike of flowers is rather grotesque, it is effective for unusual arrangements and the plant is often used in the conservatory to give a tropical effect. Zones 1 and 2. Tender. Give little moisture in spring and winter, more in summer. Well-drained, half-sunny, half-shady location. For arrangement, condition the decorative leaves in water. Flowers used extensively in Mex.

[182]

Acanthus montanus white

White Thunbergia (*Thunbergia grandiflora*)
This large flowering vine, though tender, blooms
intermittently through the warm weather. Huge
clusters of bell-shaped flowers hang in graceful
festoons and each pure white or sky-blue blossom
with its soft yellow throat is as alluring as an
orchid. Smaller-flowered, annual varieties range
in color through white and yellow to orange with
purple or brown centers. Self-seeding so it may
become a nuisance. Zones 1 and 2. Easily propagated by layering. Bank with sand for frost protection. Quick grower after frost damage. Float
in shallow bowl for effective arrangement.

Thunbergia grandiflora-v-alba – white

King's Mantle (*Thunbergia erecta*) A graceful
shrub, introduced from W. Afr., for ornamental
planting in citrus areas makes a charming dense
shrub from 3'-5' when kept pruned. Its gloxinia-
like flowers with a bright yellow curving throat
are a darker, more purplish-blue than its popular vining relative, *T. grandiflora* (color page
facing 113). They bloom continually. Its arching branches are decorated with interesting
clumps of small shiny green leaves. A white
species has similar though smaller white blossoms with yellow throat. Zones 1 and 2. Easily
propagated by cutting or layering. *T. grandiflora* hardier of the two.

Thunbergia erecta-purple, yellow

Blue-Sage (*Eranthemum nervosum*) This luxuriant herb-like shrub, originally from tropical
India, is conspicuous for its axillary spikes of
bright-blue phlox-like flowers which protrude
from green and white bracts and for its large,
dark-green embossed leaves, sometimes 8" long.
Varies in height from 2'-6'. Blooms in winter
and spring when other blue flowers are scarce.
Zones 1 and 2. Tender but withstands temperatures as low as 20°. Easy to propagate from cuttings. Excellent for shade planting from Fla. to
Calif. Coarse for arrangement but effective when
stripped of the faded blossoms and excess leaves;
keeps well.

[183]

Eranthemum nervosum-blue

Jacobinia carnea - pink

PINK-ACANTHUS (*Jacobinia carnea*) (color page facing 65) This showy conservatory plant, an herb, or sub-shrub (1'-4') from Brazil is an excellent filler for warm sections because of its large, terminal clusters of rose-colored blossoms which appear in summer and fall (winter, inside), when there is little color in the garden. The large leaves are prominently veined. Zones 1 and 2. Quick growing from cuttings. New plants are more satisfactory as the older tend to get scraggly Same culture as Bignonias, only less tender. Long lasting in arrangement.

Beloperone guttata-bracts rust-f.white

SHRIMP-BUSH (*Beloperone guttata*) (color page facing 177) This half-scandent, clambering sub-shrub with angled stems, from Mex. and tropical America, is appropriately called shrimp-bush because its conspicuous terminal bracts resemble that well-known crustacean. Small two-lipped, white flowers spotted with purple protrude from overlapping heart-shaped bracts that shade colorfully from yellow to rust. The shape of the bracts is repeated in larger deep-green leaves. A profuse bloomer throughout the year, it is often used as a house vine or pruned to a bush. Foolproof in mild climate (Zones 1-3), in full sun and summer in North. Propagates from cutting and layering. Useful and long-lasting in arrangements.

Sanchezia nobilis-yellow, red

SANCHEZIA (*Sanchezia nobilis*) There is a fascinating grotesqueness about the not particularly beautiful flowers of this Ecuadorian shrub. Transparent bright-red bracts hold spikes of waxy, yellow, tubular corollas with rims rolled under close to the stalk. Instead of falling, old blossoms dry on the square stems. This 5'-6' shrub is grown in the northern conservatories for its fine variegated foliage and should be grown more in the far South. Zone 1; risky in 2. Likes well-drained, non-acid soil and half shade. Propagates from cuttings of new wood, taken from March to July. Oddity for founda tion or border planting.

[184]

MADDER FAMILY (Rubiaceae)

FLAME-OF-THE-WOODS, JUNGLE-GERANIUM (*Ixora coccinea*) (color page facing 177) This *Ixora* flames forth in the autumn with large, flat-topped clusters, or corymbs, of vibrant, coral-red, tubular flowers but smoulders at all times in tropical sections. Closely clipped hedges are spectacular in Miami. The 4 or 5 pointed lobes gleam like sparks from the end of the long, thin corolla tubes. Prized for its glossy, emerald foliage and compact growth, it is one of the best broadleaf evergreen shrubs for ornamental planting in subtropical areas. The flowers and bark are used medicinally for bloodshot eyes and the leaves for sores and ulcers. Zones 1 and 2. Quite tender; withstands 28°. Not particular to soil, sun or shade; propagate from cutting, harder from seed. Prunes well as hedge. Lasts well in arrangement.

Ixora coccinea   orange-red

CAPE JASMINE, GARDENIA (*Gardenia Thunbergia*) This evergreen shrub may grow to a height of 10′ and bears attractive glossy, elliptic, acuminate leaves which are 4″-6″ long. The flowers are white, waxy, 3″-4″ across, and strongly scented. The calyx is split down one side and the corolla tube is 2½″-3″ long. This shrub bears a very hard, smooth, white fruit 2″-2½″ long. *G. jasminoides* is a variable evergreen shrub which grows from 2′-6′ in height. The leaves are lanceolate to obovate, acuminate, short and thick, and up to 4″ long. The often double, waxy camellia-like flower may be as large as 3″ across and is delicately fragrant. The calyx is ribbed with 5 long teeth and the cylindric corolla tube is 1″-1½″ long. The fruit is fleshy, orange in color, ribbed, and 1½″ long. Although these two Gardenias are quite similar, they may be distinguished by the following characteristics: *G. Thunbergia* bears a spathe-like calyx which is ⅓-½ the length of the corolla tube, and *G. jasminoides* has a calyx with 5 long teeth about the same length as the corolla tube. Both are planted in the open in the South as shrubs which rarely grow to the size of small trees. Zones 1-4; and Piedmont. Hardier when mulched. Withstands 15°.

Gardenia Thunbergia- white

Gardenia jasminoides - white

[185]

Pentas lanceolàta *white to red*

PENTAS (*Pentas lanceolata*) This is one of the finest, newly-introduced, free-blooming, low, compact-bedding plants for subtropical areas and for summer in the North. The attractive round clusters of star-shaped bouvardia-like flowers unfold from long tubes and bloom throughout warm weather. There are several varieties from white through the pinks and purples to dark-carmine. The Greek name refers to the 5-parted flowers and the lance-like shape of the petals and opposite leaves. A S.Afr. herb of low growth, 2'-3' tall, it is suitable for pots, for picking, and for low hedges or in front of scraggly shrubs. Zones 1 and 2; try in 3. Easily propagated from cutting. Easily grown in full sun. Keep pruned.

HONEYSUCKLE FAMILY (Caprifoliaceae)

ABELIA (*Abelia grandiflora*) The evergreen abelia is a popular shrub for foundation or hedge planting throughout the South and in Calif. During early spring and summer, it bears small, bell-shaped white to pinkish flowers in abundance. When not blooming, small bronze-green leaves and red-tinged calyces still add a note of color. Though known in Japan, India and Mex., *A. grandiflora* is a hybrid from China. Best for temperate areas (Zones 2-5). Likes sunny, well-drained, humus spots. Propagate by layering in spring; cuttings of mature wood in fall. For miniatures or low table decoration when other flowers are scarce.

Abelia grandiflora-*whitish*

COMPOSITE FAMILY (Compositae)

TRANSVAAL- or AFRICAN-DAISY (*Gerbera Jamesonii*) (color page facing 65) This brilliant "daisy" from Africa is the outstanding perennial in the subtropics for cut flowers because of its free-blooming, long-lasting qualities (over a week) and wide range of color from white through the pastel pinks and yellows to the deepest hues of orange and red. Fla. and Calif. have many outstanding new hybrids and doubles which they ship to northern markets. Zones 1-3. A "must" for all gardens to the mid-South. The crown must not be covered with earth, but a heavy mulch will prevent freezing as far n. as N.C., and with a cold frame, to N.Y. Propagate by division or slowly from seed. Divide where overcrowded. Easily arranged with its own leaves and unrivaled as a cut flower for beauty and long-lasting quality.

[186]

Gerbera Jamesonii - *many colors*

MEXICAN FLAME-VINE (*Senecio confusus*) Long cultivated in the Rio Grande section, this tropical Mex. clambering half-vine is increasing in popularity throughout Fla. for its quick growth, easy culture, and almost constant bright flowers. Though at times untidy, its brilliant, reddish-orange, daisy-like blossoms with yellow centers, borne in loose, terminal clusters, are striking against masses of thick, shiny, light-green "meaty" leaves with toothed edges. Zones 1 and 2; parts of 3. Foolproof from seed or self-layering in any soil. Heat and drought resisting. Force as an annual in Zone 3. Long lasting in arrangement.

Senecio confusus - orange-red

TARTARIAN-ASTER (*Aster tataricus*) A native of Siberia, Japan and China, this aster is frequently cultivated in this country and has escaped from cultivation in some areas. A perennial of the composite family and of rather rank growth, it often grows 6' or 7' tall and is topped with enormous clusters of bluish-purple heads of flowers in late summer and fall. The foliage and stem are hairy and the basal leaves are often a foot or more long. Zone 1, northward. Easy to grow. Propagation is by division or cutting. Suited for the back of borders or in clumps for a brilliant show. Lasts well for cutting.

Aster tataricus-purplish

MEXICAN SUNFLOWER (*Tithonia diversifolia*) The brilliant yellow-orange, sunflower-like blossoms, 5"-7" across, of this rampant-growing, rough-textured, coarse-leaved herb from Mex. brightens the border planting for a long period in late fall. It grows 8'-12' tall in informal, loose clumps, branching at the woody base. Its smaller even brighter relative, *T. rotundifolia*, the Mexican orange-daisy, is classed as an annual and should not be forgotten for fall and spring bloom, for it is foolproof. Zones 1 and 2. Easily grown from seed; an excellent filler herb for the back of the border or for a sunny dry spot where the soil is poor. Wilts if picked in the sun.

[187]

Tithonia diversifolia - orange

INDEX

(Italicized numbers refer to pages facing color plates.)

[192]

soapwort, 100; striped, 100
Gentiana autumnalis, 100
   Catesbaei, 100
   var. *nummulariaefolia,* 100
   crinita, 100; *Saponaria,* 100
   tenuifolia, 100; *villosa,* 100
Gentianaceae, 98 ff.
Gentian family, 98 ff.
George Frank, *176*
Georgia-bark tree, 128, *128*
Geranium, jungle, 185; rock, 44
Geranium maculatum, 56
Gerardia
   pectinata, 117; *purpurea,* 117
   tenuifolia, 117; *virginica, 56-57,* 116
Gerardias, purple, 117
Gerbera Jamesonii, 65, 186
Ghost-plant, 85
Giant amaryllis, 152
Giant-chickweed, 30
Giant-goldenrod, 134
Giant potato-creeper, 178
Gilia rubra, 105
Gilllenia stipulata, 47
   trifoliata, 32, 47
Ginger family, 113, 154
Ginger-lily, 154
Ginseng, 83; pineland, 138
Ginseng family, 83, 168
Girasole, 127
Glandular-skullcap, 110
Glass-worts, 29
Globe-amaranth, 29
Globe-amorette, 10
Gloriosa Rothschildiana, 81
Gloriosa superba, 81
Glory-bower *(Clerodendrum fragrans),*
   174
Glory-bower *(Clerodendrum panicula-*
   tum), 174
Glory-bush, 168
Glory of Sunninghill, *176*
Goat's-beard, *33*
Gold-crest, 20
Golden-aster, *64,* 127, 132
Golden-ball, 170
Golden-canna, 20
Golden-cassia, 54
Golden-club, 2
Golden-crest, 20
Golden dew-drops, 175
Golden-leavenworthia, 41
Golden rain-tree, 164
Goldenrod, *17, 64,* 134
Golden St. John's-wort, 73
Golden-shower, 81, *81*
Golden-star, 137
Golden-trumpet, 170, *177*
Golden-yellow star-grass, 12
Gold-in-green, 137
Gomphrena dispersa, 29
Goodeniaceae, 126
Goodenia family, 126
Gooseberry family, 46
Goosefoot family, 29
Gopher-apples, 52
Gopher-berries, 92

Gordonia Lasianthus, 74
Gourd family, 124
Gramineae, 1
Grand duke jasmine, 169
Grandsir-graybeard, 98
Granite-gooseberry, 46
Grape family, 71-72, 164
Grape-tree, 28
Grass family, 1
Grass-leaved lettuce, 143
Grass-of-Parnassus, Carolina, 45
Grass-of-Parnassus family, 45
Grass-pink, *17,* 21, 24
Grass-pink orchid, *17*
Gray, Asa, 8, 60
Graybeard, Grandsir-, 98
Gray's-lily, 8, *33*
Great-chickweed, 30
Great-laurel, 128; pink, 89; white, 89
Great-leaved magnolia, 37
Great-lobelia, 126
Green-arum, 2
Greenbriers, 13
Green-dragon, 3
Green-eyes, 137
Green-fly orchid, 23
Green-violets, 72
Grevillea robusta, 154
Gromwell, 107
Gronovius, Jan Fredrik, 143
Grossulariaceae, 46
Grossularia curvata, 46
Ground-cherry, seaside 106
Ground oak, 52
Groundsel-tree, 127, 133
Groundsels, 142
Guaiacum sanctum, 63

Habenaria, 64; Blephariglottis, 26
   ciliaris, 26; *clavellata,* 26
   flava, 26; *nivea,* 26
   odontopetala, 26; *psycodes,* 26
   repens, 26; *strictissima,* 26
Habenarias, 26
Haemodoraceae, 20
Hairy-pipeworts, 3
Hairy-ruellia, *17*
"Hairy spice-bush," 78
Hairy-vetch, 61
Halesia carolina, 95
   diptera, 95; parviflora, 95
Hamelia, 121
Hamelia patens, 121
Hardhead, *17, 64*
Hastate-leaved violet, 77
Hat-pins, 3
Hat-plant, Chinese, 175
Hawaiian hibiscus, *96*
Haw-weeds, 143
Hawthorns, 51
Heart-leafs, 124
Heart-pea, 70
Heart-seed, 70
Heath family, 86, 128
Heavenly-bamboo, 156
Hedychium coronarium, 154
Helenium, 64; tenuifolium, 141

Jamaica dogwood, 59
James grape, 72
Japanese-honeysuckle, 123
Japanese-quince, 157
Japanese-wisteria, 58
Japonica, 128
Jasmine, "African," 171
 cape, 185; Chilean, 169
 cinnamon, 169; Confederate, 171
 crepe, 172; grand duke, 169
 Madagascar, 173; "Malayan," 171
 "night-blooming," 179
 non-jasmines, 169; orange, 169
 rock, 169; star, 169
 white, 169; willow-leaved, 169
 yellow, 169
Jasminums, 98
Jasminum gracillimum, 169
 humile, 144, 169; *illicifolium,* 169
 Mesryi, 169; *nudiflorum,* 169
 officinale var. *grandiflorum,* 169
 Sambac, 169
Jean Haerens, *176*
Jeffersonia, 39
Jerusalem-artichoke, 127
Jerusalem-thorn, 97, *97*
Jessamine, Carolina yellow, 49, *49*
 star, 169; yellow, *144*
Jewel-weed, 63
Jewel-weed family, 63
Jimson-weed, 106
Joe-pye-weeds, 130
Joe-wood, 94
Joe-wood family, 94
Johnny-jump-ups, 72
Johnson, 152
Joint-wood, 28
"Jove's-fruit," 78
Judas-tree, 54
June-bean, 61
June-berry, 51
Jungle-geranium, 185
Jussiaea peruviana, 82; *scabra,* 82
Jussieu, Bernard de, 82

Kalanchoë, 161
 Daigremontiana, 112, *112*
 Fedtschenkoi, 161; *verticillata,* 161
Kalm, Pehr, 90
Kalmia angustifolia, 90
 var. *caroliniana,* 90
 cuneata, 90; *hirsuta,* 90; *latifolia, 48,* 90
King's-mantle, 183
Kiss-me-over-the-garden-gate, 28
Knotweed family, 155
Koelreuteria bipinnata, 164
 paniculata, 164
Koseteletzkya althaefolia, 72; *virginica,* 72
Kudzu-vine, 59
Kuhnistera pinnata, 57
Kurume azaleas, *176*

Labiatae, 110-11
Lachnocaulon, 3
Lactuca graminifolia, 143; *hirsuta,* 143
Lady-lupine, 55
Lady-of-the-night, 180

Lady-slipper, 21
 pink, 23; yellow, 23
Ladies'-tresses, 25
Lagerstroemia indica, 65, 165
 var. *alba, 65*
Lamb-kills, 90
Lance-leaved arrow-head, 1
Lance-leaved violet, 77
Lantana Camara, 109; *Sellowiana,* 109
Lantanas, 109
Large-flowered aster, 136
Large-flowered sabatia, 99
Large-flowered trillium, 11
Large twayblade, 24
Larkspur, Carolina, 34; Spanish, 104
Lauraceae, 78
Laurel, great, 128; mountain, 90
Laurel family, 78
Laurel-leaved smilax, 13
"Lawn-pennywort," 84
Lead-plant, 56
Lead-tree, 53
Leadwort family, 168
Leafless beaked-orchid, 25
Leather-flowers, 35
Leather-leaf grape, 72
Leatherwood, 68
Leavenworth, Dr. Melines Conklin, 41
Leavenworthia, golden, 41
Leavenworthia aurea, 41
Leguminosae, 49, 53 ff., 80, 81, **97, 113,**
 128, 158-59
Leiophyllum buxifolium, 88
 var. *Hugeri,* 88; *Lyoni,* 88
Leonotis nepetaefolia, 111
Lentibulariaceae, 118
Lettuce, 127; grass-leaved, 143
Leucaena glauca, 53
Leucothoë axillaris, 91; *editorum,* **91**
 racemosa, 91; *recurva,* 91
Liatris aspera, 132; *elegans,* 132
 gracilis, 132; *pauciflora, 64*
 pycnostachya, 129; *spicata,* 64
 squarrosa, 132; *tenuifolia,* 132
Lignumvitae, 63
Ligustrum indicum, 170
 ovalifolium, 170
"Lilac of the South, the," 165
Lilacs, 98
Liliaceae, 6 ff., 81, 112
Lilium canadense, 33; *caroliniana,* 8
 Catesbaei, 8, 64, var. *Longii,* 8
 Grayi, 8, *33; Michauxii,* 8
 superbum, 8, *33*
Lily: Amazon, 152; Atamasco, 14
 butterfly, 154; Canada, *33*
 Carolina, 8, *33;* Catesby's, 8, 150
 celestial, 15; climbing, 81, *81;* cow, 38
 Cullowhee, 14; day, tawny or orange, 9
 Eucharist, 152; fawn, 9; ginger, 154
 Gray's, 8, *33;* madonna, 152
 orange-red, *64;* pond, 38
 red spider, 152; southern red, 8
 speckled wood, 10; spider, 14, *17*
 string, 14; swamp, 14; trout, 14
 water, 38; wood, *33;* zephyr, 14
Lily family, 6 ff., 81, 112

[203]

[206]

Umbelliferae, 84-85
Umbrella-leaf, 39
Uniola latifolia, 1; *paniculata*, 1
Urechites lutea, 101
Utricularia, 42; *cornuta*, 118;
   inflata, 118; *subulata*, 118
Uvularia sessilifolia, 32

Vacciniaceae, 92
Vaccinium pallidum, 56-57
Vachellia Farnesiana, 53
Vanilla-leaf, *17*, 70
Varnish-leaf family, 70
Venus'-flytray, 42, 43
Venus's looking-glass, 125
Verbascum Blattaria, 56-57
   Thapsus, 56-57
Verbena, Carolina, 109; moss, 108
   seaside, 109; shrub, 109
Verbena caroliniana, 109
   maritima, 109; *tenuisecta*, 108
Verbenaceae (Vervain family), 108-9
Verbenaceae (Verbena family), 174-75
Verbena family, 174-75
Verbesina laciniata, 139; *occidentalis*, 139
Vernonia, 64; *angustifolia*, 130
   glauca, 130; *noveboracensis*, 129, 130
Vervain family, 108-9
Vetch, *17*, 61
Viburnum acerifolium, 33
   prunifolium, 122; *rufidulum*, 122
Viburnums, 83, 122
Vicia angustifolia, 56
   caroliniana, 61; *cracca*, *48*, 61
   grandifolia, 56; *villosa*, 61
Vinca rosea, 101
Viola arvensis, 75, 77; *canadensis*, 32, 56
   cucullata, 32; *hastata*, 77
   Kitaibeliana var. *Rafinesquii*, 75, 77
   lanceolata, 77; *Langloisii*, 32
   pallens, 77; *papilionacea*, 76
   pedata, 56, 76; *Priceana*, 76
   primulifolia, 77; *rotundifolia*, 77
   septemloba, 76; *triloba*, 76, *vittata*, 77
Violaceae, 75, 76-77
Violacea rubra, *176*
Violet, bird-foot, 76
   blue, 75; Confederate, 76
   dog-toothed, 9; green, 75
   hastate-leaved, 77; lance-leaved, 77
   meadow, 76; pale, 77
   primrose-leaved, 77; "round-leaved," 77
   seven-lobed, 76; "star," 120
   three-lobed, 76; white, *17*, 77
   wood, 76; yellow, 77
Violet family, 75, 76-77
Violet-iris, 19
Violet wood-sorrell, 62
Virginia-bluebell, 107
Virginia-creeper, 71
Virginia-rose, 50
Virginia-stuartia, 74
Virginia-willow, 46
Vitaceae, 71, 164
Vitis candicans var. *coriacea*, 72;
   Labrusca, 72; *Munsoniana*, 72
   rotundifolia, 72
Vriesia carinata hybrid Marie, 161

Wake-robins, 11
Walter, Thomas, 13, 34
Walter's-smilax, 13
Wampee, 4
Warea, 41
Warea amplexifolia, 41
Water-chinquapin, 38
Water-hemlock, 84
Water-hyacinth   (*Eichornia   crassipes*),
   16, *16*
"Water-hyacinth" (*Pontederia cordata*) 4
Water-leaf family, 105
Water-lily, blue, 38; yellow, 38
Water-lily family, 38
Watermelon, 124
Water-nymphs, 38
Water-plantain family, 1
Water-spider orchid, 26
Wax-berry, 27
Wax-flower, 173
Wax-myrtle, 27
Wax-plant, 172
Wedge-leaved haw, 51
Weeping-lantana, 109
Wells, Dr. B. W., 137
Wherry, Dr. E. T., 21
White-acanthus, 182
White-alder, 85
White-alder family, 85-86
White-arum, 2
White crape-myrtle, *65*
White-featherling, 6
White fringed-orchid, 26
White-goldenrod, 134
White great-laurel, 89
White jasmines, 169
White oleander, *65*
White-operculina, 103
White-poppy, 39
White rein-orchid, 26
White-rhododendron, 128, *128*
White-snakeroot, 130
White star-grass, 12
White-thunbergia, 183
White-topped aster, *17*, 135
White-top sedge, 1, *17*
White violet, *17*, 77
White-wood, 37
Whorled-loosestrife, 93
Wickys, 90
Wild-allamanda, 101
Wild balsam-apple, 124
Wild-bamboo, 13
Wild-beans, 60
Wild-buckwheat, 27
Wild-coffee, 121
Wild-comfry, 107
Wild-coreopsis, *17*
Wild Flowers of Louisiana (Dorman), 16
Wild-fumeroots, 40
Wild-ginger, 123
Wild-hyacinth, 9
Wild-indigoes, 55
Wild-leek, *33*
"Wild-olive," 95
Wild-pineapple, 5, *145*

Planting Zones

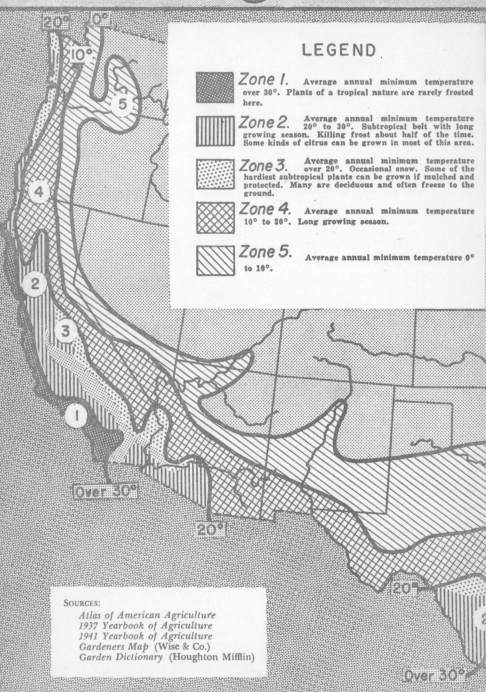

LEGEND

Zone 1. Average annual minimum temperature over 30°. Plants of a tropical nature are rarely frosted here.

Zone 2. Average annual minimum temperature 20° to 30°. Subtropical belt with long growing season. Killing frost about half of the time. Some kinds of citrus can be grown in most of this area.

Zone 3. Average annual minimum temperature over 20°. Occasional snow. Some of the hardiest subtropical plants can be grown if mulched and protected. Many are deciduous and often freeze to the ground.

Zone 4. Average annual minimum temperature 10° to 20°. Long growing season.

Zone 5. Average annual minimum temperature 0° to 10°.

SOURCES:

Atlas of American Agriculture
1937 Yearbook of Agriculture
1941 Yearbook of Agriculture
Gardeners Map (Wise & Co.)
Garden Dictionary (Houghton Mifflin)

W. K. Hubbell-52